NATO'S ANXIOUS BIRTH

ANDRÉ DE STAERCKE
and Others

NATO's Anxious Birth

The Prophetic Vision of the 1940s

Edited by Nicholas Sherwen

WITHDRAWN

ST. MARTIN'S PRESS, NEW YORK

©*Nato Review*, 1985
All rights reserved. For information, write:
St. Martin's Press, Inc. 175 Fifth Avenue, New York, NY10010
Printed in Great Britain
First published in the United States of America in 1985

ISBN 0-312-56060-5

Library of Congress Cataloging in Publication Data

De Staercke, André
  NATO's anxious birth.

  Bibliography: p.
  Includes index.
  1. North Atlantic Treaty Organization — History — Addresses, essays, lectures. I. Title.   II. Title: N.A.T.O.'s anxious birth.
UA646.3.D4   1985   355'.031'091821   84-40609
ISBN 0-312-56060-5 (U.S.)

Acknowledgement

This book is based on a series of original articles devoted to NATO's pre-history published in the *NATO Review* between June 1979 and October 1983. Thanks are due to the Editor of that publication for permission to use the articles and to the individual authors for their generous acceptance of editorial changes introduced in this volume for the sake of greater clarity.

CONTENTS

Editorial Note — *page* vii

Notes on the Authors — xi

A Landmark of the Truman Presidency — 1
The Truman Doctrine marks the end of American isolationism
CLARK M. CLIFFORD

Secret Explorations — 11
The Anglo-American initiatives
ALEXANDER RENDEL

Norway's "Atlantic Policy" — 19
The genesis of North Atlantic defence cooperation
OLAV RISTE

The Omaha Milkman — 30
The role of the United States in the negotiations
THEODORE C. ACHILLES

Searching for Security in the North — 42
Denmark's road to NATO
SVEN HENNINGSEN

The Uncertain Months — 53
British anxieties over the outcome of the negotiations
ALEXANDER RENDEL

A Change of Heart — 61
Concerns behind the discussions in France
CLAUDE DELMAS

The Pull of the Continent — 68
Portugal opts for a European as well as an Atlantic role
ALBANO NOGUEIRA

The Art of the Almost Impossible 76
Unwavering Canadian support for the emerging Atlantic Alliance
ESCOTT REID

Fate of the Nordic Option 87
The absence of guarantees for a Scandinavian defence association turns Norway firmly towards the Atlantic Alliance
GRETHE VAERNØ

Diplomacy on the Spot 95
The role of the Italian Embassy in Washington in the negotiations leading to the North Atlantic Treaty
EGIDIO ORTONA

Paul-Henri Spaak — Future Secretary General 110
Belgium sees its hopes fulfilled
BARON ROBERT ROTHSCHILD

Abandoning Neutrality 116
How and why the Netherlands joined the Atlantic Alliance
PAUL VAN CAMPEN

Opting for Commitment 133
Luxembourg consolidates its independence within NATO
NICOLAS HOMMEL

An Unarmed Nation 142
Iceland rides out the storm
OLAFUR EGILSSON

An Alliance Clamouring to be Born — Anxious to Survive 154
ANDRÉ DE STAERCKE

Principal Events, Conferences and Treaties 169

Further Reading 178

Text of the North Atlantic Treaty 183

Index 187

EDITORIAL NOTE

The North Atlantic Treaty was signed in Washington on 4 April 1949. How did it come about? What led the twelve original member-states to take this plunge into uncharted waters? Was it just one of those moments in history when necessity was the mother of invention? Undoubtedly leadership played the largest part, but if the conditions had not been right, even the vision of the leaders of the late 1940s would not have been sufficient.

Each country followed a different path to membership; each had to face internal or external problems of its own, sometimes abandoning traditional policies, sometimes in the face of heated domestic debate, before staking its future in this new enterprise.

In the following chapters, each one contributed by a distinguished commentator with special knowledge and understanding of a particular country's role in the negotiations which led up to the signing of the Treaty, the pre-history of the Alliance unfolds. By the end of these pages the reader will have been guided to Brussels, Copenhagen, The Hague, Lisbon, London, Luxembourg, Oslo, Ottawa, Paris, Reykjavik, Rome and Washington to witness the birth of the Alliance from each of these scattered vantage-points and, in the course of the story, may be struck by the continuing relevance in 1984 of the considerations on which the national decisions turned in each case, and reminded that then as now this was a voluntary process. Strenuous efforts were made to persuade the doubters, but there was no compulsion. The Western Alliance bears no comparison to the Warsaw Pact, imposed on its members and superimposed on an existing hegemonic system, whose 1955 treaty palely mimics the articles of the Washington Treaty.

The reader may be startled to discover how many of the positions adopted by the main players on the scene in 1947 and 1948 have more than a familiar ring to them: the insistence by the Americans, for example, that Europe must do more to prove itself — in 1947 — by creating and implementing the Brussels Treaty before the United States would be willing to pronounce on its own role in an Atlantic pact. Is this so different from the shot across the bows represented by the narrowly defeated Nunn proposal of 1984 and its clear message that if the United States is to repeat its guarantees with the same conviction as in the past, Europe is to assume a greater role in its own defence?

What is equally noteworthy but perhaps less obvious is that some of the earliest brushstrokes on the new canvas that was to become the Alliance have endured and assumed an importance that the artists can scarcely have foreseen. The willingness of those representing their nations to attempt to modify their own positions to enable agreement to be reached is epitomised in the anecdote recounted by Theodore Achilles: Derick Hoyer Millar announcing to the Foreign Office in London that he needed new instructions in line with the solutions agreed upon in the Working Group negotiating the Treaty, having "made his pitch and been shot down". How else could decision by consensus have survived to remain the fundamental operating principle of the Alliance? And imagine the situation today if majority wishes had been systematically imposed on reluctant minorities among the member states? The celebration of NATO's anniversaries would have been much curtailed, and with what consequences?

That there are so many parallels between 1949 and 1984 is no accident, for it is the same basic concept which has endured and thus, at least to that extent, proved itself. The security of Western Europe and the two countries of North America is one and the same; to provide it, the United States must be able to demonstrate credibly its willingness and ability to furnish not only the nuclear protection but also the reinforcements necessary for the conventional defence of Europe. In 1949 the need for refuelling facilities in Greenland, Iceland and the Azores helped to ensure recognition of the important part to be played by Denmark, Iceland and Portugal. Rapid reinforcement of Europe remains an issue no less crucial in the mid-1980s.

The common will and purpose which have given the Alliance its strength into the fourth decade of its existence are also a reflection of the broad political consensus which has sustained it nationally, as well as internationally, through periods of fundamental change. This consensus has survived the transition from pre-war to post-war values, the succession of generations, the climatic extremes of East-West relations, the leap from colonialism to emerging nationalism in much of the Third World, the advent of micro-chip technology and space-age communications, and a major shift in the global power balance. Recently there has been a new element which has threatened to succeed in undermining the roots of the Alliance where all these traumas have failed — the dilution of that broad political consensus as the expression of elected parliaments. Had this happened thirty years ago, one would have been fearful of the ability of the Alliance to survive. Today there are two factors which mitigate its effects. First, the party political manoeuvring on national security policies has had little perceptible effect on the continued overwhelming support for the Alliance of public opinion throughout the member countries. Opinion polls repeatedly confirm this. Secondly, the Alliance now has roots tough enough to withstand the temporary deprivations of occasional drought, and seasons change.

What can justify such a conclusion? Part of the answer lies in recognising the significance of the inspired doctrine introduced in March 1947 which brought an end to traditional American isolationism — the Truman Doctrine, described accurately by its author as "a milestone in history". Few historians would question the farsightedness of this policy, and few on either side of the Atlantic would deny the lessons of history by seriously advocating its abandonment. Much has been written on the subject but this book would paint a false picture indeed if it did not take President Truman's achievement as its starting point. Clark Clifford, formerly Special Counsel to President Truman, provides in his chapter the context for the book as a whole.

How else is the reader to assess the Alliance of the 1980s in the light of its antecedents? There is an extensive literature on this subject too and it is not our purpose here to provide a detailed analysis. In the final chapter, however, something of this wider perspective is to be found, for in describing the struggles of the infant Alliance, its gradual path to maturity and its early adulthood, André de Staercke looks forward as well as back and touches on the most fundamental issue of all. Was 1949 but one in a series of attempts to delay the violent course of European history? Or was it the first page in a new saga of longer-term hope and promise, in which East can meet West in mutual security and constructive dialogue? For this is where the case for Western defence rests and the future of the Alliance begins.

October 1984

NOTES ON THE AUTHORS

THEODORE C. ACHILLES was born on 29 December 1905. A graduate of Leland Stanford University and Yale, he worked as a newspaperman in California and Japan before joining the US Foreign Service in 1931. In 1947, after serving in Washington and in US missions throughout the world, he became Director for Western European Affairs at the State Department, where he was directly involved in the negotiation of the North Atlantic Treaty and in working with the Senate Foreign Relations Committee. He was Deputy Permanent Representative to NATO in London in 1950 and subsequently served on the NATO Council for several further periods between 1950 and 1960. During his distinguished career, Ambassador Achilles has held a number of additional public appointments both as the representative of his government at major international conferences, and as Personal Representative of the President. Since his retirement he has also held many private appointments, including periods as a Director and Vice-Chairman of the Atlantic Council of the United States and Governor of the Atlantic Institute for International Affairs in Paris.

PAUL VAN CAMPEN was born in Amsterdam on 15 January 1920, and took a doctorate in Political and Social Sciences at the University of Amsterdam before becoming a Counsellor in the Netherlands Association for International Affairs in The Hague in 1953. He joined NATO's Political Affairs Division in Paris in 1958. In 1966 he was appointed Special Adviser for Policy Planning under the then Secretary-General of NATO, Manlio Brosio, and in 1971 became Director of the Private Office of the newly appointed Secretary-General, Joseph Luns, an office which he held throughout Dr Luns' tenure, retiring from NATO in 1984. Paul van Campen has published a number of articles and books including *The Quest for Security: Some Aspects of Netherlands Foreign Policy, 1945 to 1950* and *The Imperator: Consequences of Frustrated Expansion*.

CLARK M. CLIFFORD was born on 25 December 1906 in Kansas. A graduate of Washington University in St. Louis and a lawyer, he served in the United States Naval Reserve from 1943 to 1946, when he was appointed Special Counsel to President Truman. He was one of the principal architects of the 1947 "National Security Act". In 1960 he served on the Committee on the Defense Establishment appointed by Senator John F. Kennedy, and represented the President-Elect in the transition period between Administrations. Under President Kennedy Mr Clifford became Chairman of the President's Foreign Intelligence Advisory Board and later advisor to both President Johnson and President Carter. He was Secretary of Defense from 1968 to 1969. Mr Clifford holds many public distinctions and academic and legal awards, and is currently Chairman of the Truman Centennial Committee.

CLAUDE DELMAS was born in 1920 and graduated from the University of Clermont-Ferrand in 1945. He began his career in secondary education and in 1956 joined the Private Office of the French Minister of Defence, M. Bourgès-Mauroury, with whom he continued to work when the latter became Président du Conseil. Subsequently he joined the staff of M. Jacques Soustelle, Governor-General of Algeria, and later that of M. Maurice Papon, Préfet de Police de Paris. In 1957 he was awarded a NATO Fellowship and then joined the NATO International Staff. For ten years he was a member of the Organisation's Information Service. M. Delmas has contributed widely to leading French newspapers and to national and international journals. He is a specialist on Atlantic and European affairs, nuclear problems and East-West relations and has published over thirty books and studies in these fields, and on the Korean war and the Cuban crisis.

OLAFUR EGILSSON was born in Reykjavik in 1936. He took a law degree at the University of Iceland in 1963 and subsequently worked as a journalist. From 1964 to 1966 he was Icelandic Regional Officer for the NATO Information Service and Secretary-General of the Icelandic Association for Western Cooperation. He then joined the Foreign Ministry, serving in the Icelandic Embassy in Paris and the Delegations to the OECD, UNESCO and the Council of Europe. In 1971 he moved to Brussels where he combined the posts of Deputy Permanent Representative to NATO with those of Deputy Head of the Icelandic Delegation to the EEC and Counsellor in his country's Embassy. Since 1974 he has held senior appointments in the Ministry of Foreign Affairs in Reykjavik and in 1983 was appointed Deputy Permanent Under-Secretary for Foreign Affairs. Ambassador Egilsson has represented Iceland at numerous international conferences, notably the 1974 Geneva Conference on Security and Cooperation in Europe (CSCE) and its follow-up meetings in Belgrade and Madrid. Since 1982 he has been Chairman of the Icelandic International Development Agency. He has published several books and articles on Icelandic and foreign affairs and holds a number of decorations and distinctions.

SVEN HENNINGSEN was born in 1910 and died in 1982. From 1953 till 1980 he was Professor of Modern History and Political Science at Copenhagen University, and Pro-Proctor of the University in 1967-9. In 1948 he was Professor of International Relations at the University of Minnesota. A member of the Danish National Commission to UNESCO from 1961 to 1970, he was on the board of a number of Danish and foreign organisations and foundations. In 1967-78 he was Chairman of the Danish Foreign Political Institute and in 1975-80 Chairman of the Atlantic Information Centre for Teachers. He published widely on foreign policy matters and on NATO.

NICOLAS HOMMEL was born in 1915, studied law in Louvain and Paris, and took up legal practice in Luxembourg in 1939. A political detainee during the war, he joined his country's diplomatic service in 1946, becoming Permanent Representative to the OECD and, from 1950 to 1959, Luxembourg's

Ambassador to NATO. During a distinguished diplomatic career he has been Secretary-General of the Foreign Ministry in Luxembourg and has represented his country as Ambassador in Brussels, Paris, London and Bonn. From 1973 until his retirement in 1980 Nicolas Hommel was Secretary-General of the Council of the European Communities. He holds many national and foreign distinctions and decorations.

ALBANO NOGUEIRA joined the Portuguese diplomatic service in 1941 at the age of 30 and served in the Portuguese Embassy in Washington, Pretoria, Tokyo, London and at the United Nations, and in the Consulate at Bombay. From 1961 to 1964 he was Director-General of Economic Affairs in Lisbon, subsequently becoming, in turn, Ambassador to the European Communities and to NATO. After two years as Ambassador in London, he returned to Lisbon in 1977 as Secretary-General in the Ministry of Foreign Affairs. He was Visiting Professor in International Relations at the University of Minho, Braga, from 1978 to 1980. Since his retirement Ambassador Nogueira has been a contributor to several leading Portuguese newspapers.

EGIDIO ORTONA, born in September 1910, is a graduate of the Universities of Grenoble and Poitiers and has a doctorate in law from the University of Turin. He entered the Italian diplomatic service in 1932 and has had a large number of consular and diplomatic assignments in Europe, the Middle East, South Africa and the United States. In 1958 he was appointed Ambassador to the United Nations where for two years he was President of the Security Council. He was subsequently Secretary-General to the Ministry of Foreign Affairs in Rome and served as Italian Ambassador to Washington from 1967 to 1975. Egidio Ortona has been Chairman and president of numerous public and private organisations and is on the Board of Directors of several large companies. He is a Governor of the Atlantic Institute and Chairman of the Consultative Council of the Bologna Centre of Johns Hopkins University, and a Vice-Chairman of the Trilateral Commission. He has received a number of Italian and foreign decorations for his services to public life.

ESCOTT REID was born in Ontario, Canada, in 1905. A graduate of Toronto and Oxford Universities (a Rhodes Scholar at the latter), he was national secretary of the Canadian Institute of International Affairs in 1932–8 and later Acting Professor of Government and Political Science at Dalhousie University, Halifax. In 1939 he joined the Department of External Affairs in Ottawa, and went on to hold a number of important public appointments, becoming Deputy Under-Secretary of State for External Affairs, High Commissioner to India, Ambassador to West Germany, Director of the South Asian and Middle Eastern Departments of the World Bank, and Principal of Glendon College, Toronto. He has been a member of the Canadian delegations to a large number of international conferences and in 1970–2 was Consultant to the Canadian International Development Agency. He holds several honorary doctorates and public awards. Escott Reid is the author of four books including *Time of Fear and Hope: The Making of the North Atlantic Treaty* (1977).

ALEXANDER RENDEL was born on 14 May 1910, and practised as a solicitor after graduating from Oxford University. In the 1939–45 war he served as an artillery officer in the United Kingdom, Middle East and Crete and, from 1946 to 1949, during the crucial period preceding the signing of the Treaty of Washington, was on the staff of the British Embassy in Athens. Mr Rendel was for many years (1949–76) diplomatic correspondent for *The Times*, writing extensively on NATO and the Western European Union and on South East Europe and Cyprus. He was awarded the CBE in 1975.

OLAV RISTE graduated from the University of Oslo in 1959 and took his doctorate at Oxford before joining the Office of History of the Norwegian Defence High Command. He has been, *inter alia*, a Research Fellow in American History at Harvard University, a Visiting Scholar at the London School of Economics and a Visiting Professor at the Free University of Berlin, and is currently Director of the Research Centre for Defence History at the Norwegian National Defence College as well as Associate Professor of History at the University of Bergen. A Fellow of the Norwegian Academy of Science and Letters, Dr Riste is the author of published works and articles on the history of Norwegian foreign and security policy during and between the World Wars and during the early years of the North Atlantic Alliance, the most notable being *"London-Regeringa": Norge i Krigsalliansen 1940–1945*, a history of the Norwegian government-in-exile described in *International Affairs* as "a formidable achievement".

BARON ROBERT ROTHSCHILD was born in December 1911. He took a doctorate in Political Science at Brussels University and in 1937 joined the Ministry of Foreign Affairs where he worked in the private office of Paul-Henri Spaak in 1937–8. Joining the Belgian army as a reserve officer in 1939, he was made a prisoner-of-war but freed in 1941, after which he escaped from occupied territory and joined the Belgian government-in-exile in London. From 1941 he served in Belgian embassies in Europe, the Far East and the United States and in 1952–4 was Deputy Belgian Permanent Representative to NATO. In 1954 he rejoined the private office of Foreign Minister Spaak, until he was appointed Permanent Representative to the Interim Committee of the EEC in 1957. In 1958–60 he was Ambassador in Belgrade and in 1960 went to Katanga as Head of the Belgian Mission, returning in 1961 to head M. Spaak's Private Office. He then became Ambassador to Switzerland, and later Chairman of the Executive Committee of GATT, Ambassador to Paris and Ambassador in London. Baron Rothschild has published books and many articles on international affairs.

ANDRÉ DE STAERCKE was born in Ghent on 10 November 1913, studied in Belgium and France and took doctorates in law at the Universities of Louvain and Paris. Entering public service in 1939, he became Director of the Private Office of the Prime Minister and Secretary of the Cabinet-in-exile in London in 1943. He subsequently became head of the Private Office of the Prince Regent of Belgium. In 1950 he entered upon a long and distinguished career in NATO, becoming Permanent Representative of Belgium from 1952 until his retirement in 1976. For most of this period André de Staercke was *Doyen* of the North Atlantic Council and as such played a particularly significant role in the affairs of the Alliance at crucial stages in its development. He became one of the Alliance's most distinguished diplomats and acknowledged experts, and was responsible for many of the initiatives which have enabled the NATO Council to develop its consultative function. Ambassador de Staercke has received a large number of public distinctions and foreign and national decorations, and in private life holds directorships and managerial positions in several important companies.

GRETHE VAERNØ was born in October 1938. After graduating from Oslo University in 1969, she became a freelance writer on foreign policy and international affairs, writing for *Aftenposten* and leading journals. In 1971 she joined the Press Section of the Norwegian National Council of Women and was President of the Council in 1979–81. She has been a member of several national and international commissions in the field of human rights, and both a student of the Norwegian National Defence College and a NATO Fellow. Elected to the Norwegian Parliament in 1981, Grethe Vaernø is a member of its Standing Committee on Foreign Relations and is Vice-Chairman of the Norwegian European Movement.

A LANDMARK OF THE TRUMAN PRESIDENCY

The Truman Doctrine marks the end of American isolationism

Clark M. Clifford

When Franklin D. Roosevelt suddenly died on 12 April 1945, Vice-President Harry S. Truman was thrust into the White House at one of the great turning points in modern history. The Second World War was at a climax. In Europe, the Red Army was fighting on the approaches to Berlin. In the Far East, United States air, sea, and ground forces were moving relentlessly toward the Japanese home islands. The world stood on the threshold of profound change, in which the United States was certain to exert major influence. And in a flash of drama, late on an April afternoon, the conduct of American foreign policy passed into the hands of a succeeding vice-president, only eighty-three days in office, about whose views the American people at large, and the wartime allies of the United States, knew next to nothing.

In the excitement, it was not forgotten that at roughly a comparable point at the close of the First World War the American people had reverted to the isolationism that had had a long history in the United States. Although Woodrow Wilson had fought to exhaustion to make the United States a member of the League of Nations, the Senate had rejected his proposal. A quarter of a century later, in the midst of the Second World War, President Roosevelt proposed that the United States become a member of the United Nations, in effect a successor to the League. But he died even before the conference to organize the United Nations was held in San Francisco and, of course, before the issue of American membership could be submitted to the Senate.

That question and all the momentous problems of United States relations with the new order that was to emerge in Europe and Asia were now suddenly thrust upon Harry Truman, who had not had a shred of experience in the conduct of foreign relations. In fact, the only time he had ever travelled abroad was in uniform on a troopship in 1918. In ten years, as a Senator, his main concerns had been domestic affairs and, after Pearl Harbour, the efficiency of the national defence effort in the United States. The stronghold of traditional isolationism in America was the Midwest and Truman came from Missouri.

In the shock of Roosevelt's death, political leaders everywhere, and none more so than British Prime Minister Winston S. Churchill, were in an anguish of doubt about where the new President of the United States stood on post-war foreign relations. After the Roosevelt funeral, Churchill sent British Foreign Secretary Anthony Eden and Lord Halifax, the British Ambassador in Washington, to talk to President Truman. Later, Eden cabled Churchill: "I believe we shall have in him a loyal collaborator, and I am much heartened by this first conversation."

Eden's observation was an accurate one. Truman was from the Midwest, but he was no isolationist. As a farmer in Missouri in 1912 he had come under the influence of Woodrow Wilson and later he was wholeheartedly in favour of America joining the League of Nations. For decades, even after he became President, Truman carried in his wallet a paper on which were written lines from Tennyson's *Locksley Hall*, ending:

> *Till the war-drum throbb'd no longer,*
> *And the battle flags were furled*
> *In the Parliament of man,*
> *The Federation of the world.*

In a little-noted speech in the Senate in 1943, the year before the Normandy invasion, Senator Truman said: "I am just as sure as I can be that this World War is the result of the 1919–1920 isolationist attitude, and I am equally sure that another and worse war will follow this one, unless the United Nations and their allies and all the other sovereign nations decide to work together for peace as they are working together for victory. I want this greatest of legislative bodies to go on record in no uncertain terms that it will not again contribute to a condition that will cause another world war."

Almost from the start of his presidency, it was clear that in his world outlook Truman stood in the tradition of Woodrow

Wilson and Franklin Roosevelt. Truman was convinced that the old isolationism was an anachronism and worse — that for the United States to turn its back on the world, if that were possible, would be to prepare the soil for the seeds of future war. Drawing upon the so-called lesson of Munich, he believed that the United States should take a positive stand with like-minded allies to enforce the peace and to establish a system of collective security to restrain and, if need be, punish aggression. Following in the footsteps of Wilson and Roosevelt, he was determined that no power hostile to the United States and hostile to democratic institutions should be allowed to dominate Europe.

Truman's first decision as President was to announce that the United Nations organisational conference in San Francisco should proceed as scheduled despite Roosevelt's death. After the Charter was signed, Truman went before the Senate personally and appealed for ratification. It carried with only two dissenting votes. His hopes that in the long run the United Nations might play a valuable role in preserving peace never blinded Truman to the difficulties of maintaining harmony through a conclave of diverse sovereign states. And difficulties arose all too soon. Even in Roosevelt's last weeks in office, serious differences had developed with Stalin over the terms of the Yalta Agreement regarding the future status of Poland. By the time Truman assumed the conduct of foreign policy, relations between the Soviet Union and the Western Allies were deteriorating gravely in the crosscurrents of deep and powerful historical forces.

Soviet actions in Poland angered the new President, but he persisted in believing that the conflicts could be resolved. Even after his trying sessions with Stalin at the Potsdam Conference in August 1945, Truman continued to feel that he and Stalin could find a way through their differences. It was not to be. The opposing nature of American and Soviet aims and interests constantly heightened tensions. By 1946 the dreams of VE-day had dissolved into the Cold War.

In Washington's eyes, certainly, the deterioration in Soviet-American relations had become relentless. Piece by piece, the Soviets were imposing their authority on half of Europe, contrary to the United Nations Charter and the principle of self-determination. Abetted by the Kremlin, local Communist Parties were seeking to take over war-ravaged Italy and France.

Against Truman's adamant opposition, Stalin was angling for a share in the occupation of Japan. The American people were shocked by the revelation of a Soviet spy-ring operating in Ottawa for the purpose, among other things, of penetrating atomic secrets in the United States. In the tradition of the Tsars, Stalin was pressing Turkey for the right to share in the defences and control of the Black Sea Straits. In Greece, leftwing guerrillas were challenging the government. At the same time, Moscow was pressing to extend its influence in Iran and hence in the Middle East by supporting leftist political elements striving to separate Iran's northernmost province of Azerbaijan from the Iranian nation. Stalin made demands for German reparations that went beyond anything Truman would accept. Through a long series of events, Soviets and Americans moved into a position of rivalry for control of defeated Germany. In the United Nations, the Soviets rejected the United States proposal, known as the Baruch Plan, for future international control of nuclear energy.

By early 1947 an international crisis had developed as a result of the conflicts between the Soviet Union and the Western powers. At the time I was Special Counsel to President Truman. Among my principal duties was working on major presidential speeches. Since many of those speeches dealt with foreign affairs, I was necessarily involved in discussions of foreign policy and often worked closely with Dean Acheson, then Under Secretary of State at the right hand of Secretary of State George C. Marshall. Because Marshall was forced to spend a great deal of time attending the numerous post-war conferences outside Washington, Acheson was Acting Secretary for long stretches and was already on the way to becoming the chief architect of American foreign policy, as was clearly the case after he became Secretary of State in 1949, when Marshall stepped down because of illness.

As nearly as I can recollect, it was on 21 February 1947 or perhaps the next day that I heard, in the White House, that President Truman was faced with a grave decision regarding Greece and Turkey. What had happened was that on that day the British delivered to the State Department what is called in diplomatic parlance "a blue piece of paper". The term denotes an important formal message. The *aide-mémoire* was in two parts. One part said: "His Majesty's Government have already strained their resources to the utmost to help Greece and have granted . . . assistance up to 31 March, 1947. . . . The United States Government will readily understand that His Majesty's Government, in view of their own situation, find it impossible to grant further financial assistance to Greece." The crux of the other part was: "In their existing financial situation, His Majesty's Government could not, as the United States will readily appreciate, contemplate themselves making any further credits available to Turkey."

The drain of two world wars had cost Great Britain its empire. The question now thrust at President Truman was whether the United States should assume the historic British role in the Eastern Mediterranean or risk seeing Greece and Turkey fall under Soviet domination, with incalculable consequences for the free world.

I attended a meeting at which the President, with great earnestness, discussed the meaning of the *aide-mémoire*. He talked about the blood and treasure the United States had lost in the Second World War. Now, he said in so many words, the country was again faced with the possible danger of further conflict. He noted that the Eastern Mediterranean had traditionally been in Britain's sphere and that the United States could decline to take up the British burden in Greece and Turkey. He stated he had no funds available for such an undertaking. He would have to ask the economy-conscious Eightieth Congress for an appropriation. But he felt that he had to demonstrate to Stalin that the United States was determined to defend its interests. For the moment he held in abeyance the question posed by the British.

Then, on 26 February, the Secretaries of State, War, and Navy recommended that "we should take immediate steps to extend all possible aid to Greece and, on a lesser scale, to Turkey." Truman promptly approved the memorandum in principle. While this was indeed a major decision, it represented no radical change in policy. For since the start of 1946, Truman had been moving in various ways to contain Soviet expansionism, and aid to Greece and Turkey was another step in that direction, though a far-reaching one. When consulted, leaders of the Republican-controlled Eightieth Congress, notably Senator

Arthur H. Vandenberg of Michigan, Chairman of the Senate Foreign Relations Committee, made it clear that Truman would have to make a powerful statement on the Soviet menace to win a $400 million appropriation needed for Greece and Turkey. That was an important factor in the strong tone of the address that Truman delivered personally to a joint session of Congress on 12 March 1947.

Spadework on the speech was done in the State Department under Acheson's eye. Our task in the White House was adapting the material for Truman's style of delivery. The speech quickly became known as the Truman Doctrine because of a statement in which the President spoke these words: "I believe that it must be the policy of the United States to support free peoples who are resisting attempted aggression by armed minorities or by outside pressures."

Afterward, the Truman Doctrine was criticised, especially by revisionist historians, for imbuing American foreign policy with an interventionist spirit that contributed to our involvement in the Korean and Vietnam wars. Unquestionably Truman's words were strong and did provide a certain motif in American foreign relations. But Truman did not intervene indiscriminately. Notably, he did not send American armed forces on a large scale into China to try to thwart the Communist take-over. Nor were his words binding on his successors.

Congress voted the money for Greek-Turkish aid. Greece and Turkey never came under Soviet hegemony. On the contrary, they became, in 1952, signatories of the North Atlantic Treaty.

The Truman Doctrine had a significance beyond its immediate objective in the Eastern Mediterranean. Acheson later said that when the President decided to take a strong stand in Greece and Turkey, it was a signal for the Administration to go ahead in this direction, meaning, evidently, with the support and defence of Western Europe.

After the Second World War, the perennial crucial issue in Europe was Germany. The Potsdam Conference decided in 1945 in favour of joint control by the United States, Great Britain and France, which had zones of occupation in the western half of Germany, and by the Soviet Union, which occupied the eastern half. Steadily, and not always because of Soviet actions, to be sure, four-power rule fell apart. A dangerous rivalry developed between the Allied occupying powers in western Germany and the Soviet occupation in the eastern part of the country.

In the precarious circumstances, the occupying powers met in London in December 1947, but efforts to restore four-power rule collapsed. As a result, a conviction grew in Great Britain, France, the United States and the Benelux countries (Belgium, the Netherlands and Luxembourg) that rearmament would be necessary for the security of Western Europe. Presently, Ernest Bevin, then the British Foreign Secretary, proposed to Paris that the British and French jointly offer a mutual defence treaty to the Benelux countries as a first step toward a wider defence of the North Atlantic community. Secretary Marshall formally expressed American willingness to support a Western European Union, such as was proposed by Bevin. Negotiations had just been completed when the West was stunned by the news that the Communists had seized power in Czechoslovakia. In all the Western capitals the coup was seen as a direct challenge by Stalin. The West responded with two notable actions on 17 March 1948. In Brussels Britain, France, Belgium, the Netherlands, and Luxembourg signed a fifty-year collective defence treaty. In Washington President Truman went before a joint session of Congress to ask for a full restoration of selective service, which had expired, and he declared that the Brussels Pact deserved the full support of the United States "by appropriate means".

No one who lived through it will ever forget the feverish activity in Washington in the spring and summer of 1948, a time, incidentally, when President Truman was also preparing for what was generally considered to be a futile campaign for re-election in November. In the wake of the Communist seizure of Czechoslovakia, the Soviets blockaded Berlin, a dangerous move that President Truman successfully countered with an airlift. Legislation authorising the Marshall Plan was moving through Congress. And the Vandenberg Resolution, in effect giving the Administration clearance to negotiate the North Atlantic Treaty, subject to later Senate approval of the document, was adopted. The treaty negotiations progressed briskly, but because of Congressional preoccupation with the Marshall Plan and the foreshortened session due to the political campaign, the NATO treaty was not signed and submitted to the Senate until 1949.

During the negotiations, Senator Vandenberg cautioned the

Administration that the Senate would not consent to a treaty that automatically committed the United States to go to war. It was made clear, therefore, to other representatives in the negotiations that the treaty must recognize that, in the United States, Congress alone had the power to declare war.

The treaty was signed in Washington on 4 April 1949. At a state dinner for the assembled diplomats, President Truman offered a toast. "I think", he said, "we have really passed a milestone in history today. . ." Another way of putting it would have been to say that the treaty, under which the United States, Canada and ten European signatories resolved "to unite their efforts for collective defence", marked the end of the old American isolationism. The United States had followed a policy of no permanent alliances outside the Western Hemisphere practically since George Washington's Farewell Address on 17 September 1796.

"It is a simple document," Mr. Truman said at the signing ceremony, "but if it had existed in 1914 and 1939, supported by the nations who are represented here today, I believe it would have prevented the acts of aggression which led to two world wars."

The President was convinced that, once the treaty was in force, an aggressor would be deterred from picking off one nation after another, as Hitler had done, until all Western civilisation was in peril. The deterrence, he felt, was posed in the critical Article 5, which says that "an armed attack against one or more of them in Europe or North America shall be considered an attack against them all." On 21 July 1949, the Senate advised ratification by a vote of 82 to 13.

A mere decade earlier such an act by the United States Senate would have been unimaginable. It is also true that when the North Atlantic Treaty was ratified in 1949, few envisioned anything like the vast institution of the North Atlantic Treaty Organisation — NATO — that we know today. That was the case even though, as a companion to the Treaty, Congress passed the Mutual Defense Assistance Program for what would now seem a modest start on the rearmament of the other signatories.

In the main, a combination of two momentous events caused the burgeoning of NATO. One was the Soviet breakthrough in the field of nuclear weapons in September 1949, ending the

American monopoly. The other was the outbreak of the Korean war in June 1950. The Soviet nuclear breakthrough led President Truman not only to approve the production of the hydrogen bomb but to order a reappraisal of the American defence posture. As expounded in an extraordinary top-secret document, NSC 68, the reappraisal contemplated an expansion of the annual defence budget from $13.5 billion to $40 billion or more. While Truman was pondering the feasibility of proposing such a drastic programme, North Korea invaded South Korea. After committing the United States to what proved to be a major war in Korea, the President found it necessary to accept the recommendations of NSC 68, and Congress went along with huge new appropriations, a sizable part of the money earmarked for speeding up the rearmament of Western Europe.

A further effect of the North Korean attack was to cause Truman and Acheson great anxiety as to whether, if the Soviets were to march in Europe, West Germany would prove to be as vulnerable as South Korea had been. Truman decided, therefore, on a number of initiatives. One, unprecedented in American history, was to deploy more troops in Europe as part of a collective force at a time of peace on the continent. Another was to rush the establishment of an independent West German government, destined (in 1955) for membership in NATO. A third, which did not come to fruition until the Eisenhower Administration, was to rearm West Germany. NATO grew dramatically.

Today the North Atlantic Treaty still remains the cornerstone of American policy in Europe. NATO stands as a great landmark of the Truman presidency. The Alliance has been through many crises, partly because the exorbitant cost of modern weapons has made it difficult for members to reach military goals. But just when disintegration threatens, NATO springs to new life. Western Europe and hence the United States have been more secure as a result of it. NATO survives because it serves a vital purpose and will continue to do so. At the time the Treaty was negotiated, Acheson advised Truman that "in the world of today, the security of the parties . . . is so interdependent that an armed attack on any one of them would be, in effect, an attack on all."

In other words, even if there were no NATO, we would have to act as though there were, if Europe were attacked from the East. The fact that there has been no war in Europe in the thirty-five years in which the Treaty has been in effect is, I believe, a testament to its deterrent influence.

SECRET EXPLORATIONS
The Anglo-American initiatives

Alexander Rendel

Thirty-five years on, it is difficult for those who have grown accustomed to the existence of NATO to recall or imagine the uncertainties and lack of organization in Western defences, which prevailed in the years immediately following World War II, before NATO was established. During the last few years, however, there has been an annual and salutary reminder, as the confidential British Foreign Office documents of 30 years before have been published; documents for the year 1948 give a particularly poignant reminder, since they cover the anxious period of the Communist seizure of Czechoslovakia in February and the Berlin blockade which followed in July.

It is clear from the Foreign Office papers that the problem of the defence of West Europe against Russia was uppermost in the minds of the most senior Foreign Office officials throughout the year, but the situation in 1948 was paradoxical. Three years after the German surrender, Germany was still, in the view of many millions of Europeans, a near-enemy and potentially the most likely enemy of the future, whereas Russia, which was by then blocking and disrupting Western policies on all fronts, was still widely regarded as an ally who had valiantly and with terrific suffering borne the brunt of Nazi aggression.

Thus we find Ernest Bevin, the British Foreign Secretary, working hard from the start of the year to establish the Western Union between Britain, France and the Benelux countries, avowedly to prevent any renewal of German aggression, but taking into account the need both for protection against Germany and, at the same time, for a rapprochement between Germany and her western neighbours — a two-sided (some would say two-faced) policy which would help to reassure and steady, in particular, France on the score of Germany, while being in fact essential as a defence against Russia.

The Western Union was indeed about to become a foundation

stone upon and around which the NATO alliance was built (with Germany within a few years as a leading member of it). The way Bevin's mind worked can be seen from a note sent on 1 March 1948 to Clement Attlee, the Labour Prime Minister.

Bevin wrote as follows: "Instead of being bottled up in Central Europe, we feel the Germans have a great contribution to make to the world's industrial and social development. Our aim is to protect ourselves against any further aggression by Germany and at the same time to bring her back into the comity of nations as a united entity on a democratic basis, with democracy as Western civilisation understands it. In this connection, of course, you must not forget the French. We all talk too much about Germany. Our approach, therefore, to a reorganisation of economic, social and defence weapons is a good neighbourly policy, first with the French and now with Benelux. In view of the fact that France has been invaded so many times and paid such a price, we must therefore arrange our defences and our responsibilities to give the French the assurance of her security as far as we humanly can. Nothing we shall do in the Western Union will be directed against Russia or any other country in the world, but it is an essential development both economically, socially and defensively."

Bevin's detractors will say that this note in its final rather blurred reference to the need for defence against Russia is hypocritical; others that Bevin was in fact — even though gropingly — doing his practical best for democracy "as Western civilisation understands it".

At the beginning of March 1948 Western Europe had just been suddenly and forcefully alarmed by the Communist seizure of Czechoslovakia on 25 February 1948, and at about this time reports were circulating that the Soviet Government, which had brought heavy pressure to bear on Finland, was about to apply similar pressure upon Norway. Greek and Turkish diplomats were also making clear their Governments' nervousness.

Bevin was in no doubt that concerted Western action was needed to stop the rot, as a telegram sent to Washington on 10 March clearly shows — but it also shows how inchoate were any plans for the organisation of Western defences. After referring to the probability of an imminent threat to Norway, and to instructions sent to British representatives in Oslo to urge

Norway not "to put her foot on the slippery slope by sacrificing her right to conclude pacts with whomsoever she chooses", Bevin's message continued: "This may not be enough. Nor can we afford at this moment to risk a Norwegian defection which would involve the approach of Russia on to the Atlantic and the collapse of the whole Scandinavian system. This would in turn prejudice our chance of calling any halt to the relentless advance of Russia into Western Europe."

There were, as Bevin saw it, two threats: an extension of the Russian sphere to the Atlantic, and a political threat to destroy all the effort made by the British Government (with full United States approval) to build up a Western union.

Bevin's move was clearly intended to drive home to the United States Government the urgent need for a bold step. "I have", he wrote, "for some time been turning over in my mind how best to tackle the problem which has been brought to a head by the impending Russian move on Norway." "Very early steps", he suggests, "before Norway goes under", should be taken to conclude "a Regional Atlantic Approaches pact of mutual assistance in which all the countries directly threatened by a Russian move to the Atlantic could participate, for instance the United States, the United Kingdom, Canada, Ireland, Iceland, Norway, Denmark, Portugal, France, and Spain when it has a democratic regime."

Bevin then explained that after long careful thought he had decided against inviting the Scandinavian countries to join in the negotiations for a Western union, because Britain, France and the Benelux countries could not by themselves effectively defend Scandinavia and because there was not "quite the same outlook from France and the Benelux countries as from the Scandinavian countries in regard to the whole problem of Atlantic security".

Practical steps to meet the danger to Norway (and possible consequent encirclement of Sweden) were, however, needed at once and Bevin proposed as the most practical course the following three systems:

(1) the United Kingdom, France, Benelux system with United States backing;
(2) a scheme of Atlantic security with which the United States would be even more closely concerned; and

(3) a Mediterranean system which would particularly affect Italy.

"We are pressing ahead", Bevin wrote, "for the first system, but in view of the threat to Norway, the Atlantic security system is now even more important and urgent. I am convinced, therefore, that we should study without any delay the establishment of such an Atlantic security system, so that if the threat to Norway should develop, we could at once inspire the confidence to consolidate the West against Soviet infiltration, and at the same time inspire the Soviet Government with sufficient respect for the West to remove temptation from them and so ensure a long period of peace. We can turn the whole world away from war if the rest of the nations outside the Soviet system become really organised, and in turn save Russia herself. Please put all these considerations to the Secretary of State and suggest that as a preliminary we should very secretly explore my proposal."

Bevin added that a similar proposal was being made to the Canadians and hoped that the United States Government would agree to Canadian participation. He did not wish to alarm the United States or ask for an immediate reply, but urgent action was needed, and a prompt response would be helpful, particularly on the United States attitude to the Norwegian problem, since the Norwegian Foreign Minister had asked to see Bevin on 15 March, when they would both be in Paris. Meanwhile on 9 March Bevin had sent a message to the British representatives at the Western Union negotiations saying: "You should know that my mind is moving towards dealing with Scandinavia by means of another treaty system."

The American response came promptly indeed. George Marshall decided not to discuss Bevin's proposal in the United States Cabinet, but to see President Truman alone, "fearing leakage", as his aide told the British Embassy, "if other members of the Cabinet are admitted to the secret." But within 48 hours he had replied to the British Government: "We are prepared to proceed at once in the joint discussion on the establishment of an Atlantic Security system. I suggest the prompt arrival of the British representatives next week."

Bevin replied on 14 March, warmly welcoming Marshall's response. On 15 March Sir Orme Sargent, the Permanent Under-Secretary at the Foreign Office, reported to Bevin, who was in Paris, that the British Cabinet would be considering the proposed "Atlantic Pact" on 16 March, and suggesting the lines on which Gladwyn Jebb (later Lord Gladwyn) should present the British point of view at the talks in Washington. Events were obviously moving fast. The Brussels Treaty was signed on 17 March.

It is particularly clear from the instructions proposed for Jebb how much uncertainty still surrounded the form which the eventual Western security system was to take. Jebb, it was suggested, should first point to the Brussels Treaty and get the Americans to disclose how they viewed backing it up, or even acceding to it (from subsequent messages it is clear that the British did not really hope for United States accession, but would have been well pleased to have it). Secondly, Jebb should focus discussions on the proposed Atlantic Pact. "From our point of view one reason favouring a separate Atlantic system, which would, however, include ourselves and perhaps France, is that Benelux might possibly be reluctant to assume additional commitments by inviting, for instance, Norway to join. Moreover, since the proposed Atlantic system would be for defence only, it might be better to avoid the economic difficulties which might arise if non-European states acceded to the Brussels Treaty." Thirdly, Jebb should suggest that the Pact be initially confined to the United States, the United Kingdom, and Canada, leaving the way open for subsequent accession by European states with an Atlantic seaboard, "though naturally Spain and Germany could only join in circumstances very different from the present." "It would seem", the telegram goes on, "absolutely necessary that as many of the countries of the Atlantic seaboard as possible, more particularly Norway, should accede to it. . . ." Jebb might also suggest a formula for the geographic area of the Treaty, which would extend from "30 degrees north latitude and include all countries with an Atlantic littoral, which could be deemed to be possessed by Norway, Sweden, Denmark (including Greenland), Germany, Holland, Belgium, France, Spain, Portugal, Iceland and Ireland."

Finally, "as regards the possible Mediterranean Pact our feeling here, is strongly to the effect that it will be far better to associate Italy with the Brussels Treaty as soon as we can. [. . .] Italy likes to be considered as a Western power and would hardly appreciate being lumped together with Greece, Turkey, and even Egypt." If Italy did accede to the Brussels Treaty, there would really be no need for anything else in the Mediterranean though the possibility of some Middle East regional pact was not out of the question. The practicability of an eventual Mediterranean system could, however, be explored.

Thus the Washington talks which were to lead to the establishment of NATO in April 1949 began in unseemly haste (and remarkable secrecy), with no clear plan in view as to how the European agreement on defence of the central front should be backed or joined by the United States, or how defence of the flanks of the central front should be organised or linked with the centre. Throughout the year, however, with Communist pressure continuing, the most senior men in the Foreign Office were constantly at pains to provide an up-to-date report on the situation in Europe for presentation to the Americans as opportunity demanded — first, in view of the continuing threat to Greece, then in face of the continuing reports of imminent pressure on Norway, and throughout the second half of the year to support the need for an effective Western defence system as the outcome of the discussions which began in Washington in July between the United States, Canada, Britain and the four other Western Union members. Whether the main Foreign Office report, known as "Bastion", should also be shown to the French was the subject of several exchanges within the Foreign Office, where it was eventually decided that the desirability of the closest possible cooperation with France was outweighed by the risk that confidential Western views would be leaked to Russia if a document was presented. It was eventually agreed that only points from the report could be discussed informally with friendly senior French officials.

The gist of "Bastion" and other similar preparatory reports was summarised on 8 October 1948, in note form, by Sir Ivone Kirkpatrick, later Permanent Under-Secretary at the Foreign Office, for use by Bevin in speaking to Commonwealth Prime Ministers. The Memorandum is occasionally annotated by Bevin himself, whose "fist" — rightly so called rather than "hand" — betrays his total lack of formal education and in spite of its illegibility conveys also the inescapable conviction that the man himself was a tower of strength. The world,

Kirkpatrick declares, is being divided remorselessly by the Soviet Government's actions into two camps, which are already in collision in Europe, the Middle East, Asia and South America. The free nations must, therefore, get together. Collective arrangements — "admittedly a second best" — were the only alternative. In Europe, the Russians, having consolidated their grip on East Europe, were seeking to infiltrate West and Southern Europe, probing the Western line at possible weak points — Germany, Austria, Trieste, Greece and Turkey — with Communists also active in all Western countries, particularly France and Italy. Perseverance would be required since the Soviet Government would not abandon its aims at the first sign of effective opposition.

In Germany, in the light of the Berlin airlift and other measures, Sir Ivone Kirkpatrick informed Bevin that "most of the best cards are in our hands." In Austria and Trieste, the Allies had to hang on until there was a change of attitude on the Communist side. Greece was needed, because her loss would turn Turkey's flank, weaken Italy's strategic position, and threaten communications through the Mediterranean. Long-range aid would be needed for years ahead. Turkey was a static but solid barrier, handicapped, however, by a weak economy and the need to keep large forces mobilised. In Western Europe the Swedish elections had strongly confirmed the strength of Swedish neutralism. Norway and Denmark were more disposed to enter a Western group and receive Western support against Russia, but they strongly desired Scandinavian solidarity. Their governments, therefore, needed some time. In France a Christian Democrat-Socialist alliance was held together only by fear of De Gaulle, and the outlook was disquieting. In Italy, the Communists and their fellow-travellers had lost the elections (though polling 8 million votes). The Government (de Gasperi) wanted to align Italy with the West, but had to move cautiously. Italy would be a liability. If no provision was made for Greece and Turkey, there would be protests from them.

Kirkpatrick's memorandum then added that the British Government had meanwhile completed the Western Union treaty with the "hard core" of West Europe; was seeking United States participation through the meetings in Washington in which all the Western Union countries had joined; and hoped to extend the circle of partners to include "other European countries" (which would possibly be Norway, Denmark, Italy, Portugal, Ireland and Iceland). He concluded that the third stage of extending the partnership should be handled step by step without attempting to lay down a constitution. He noted the desire for European federation voiced by many European politicians, including Frenchmen, at The Hague conference of the European Movement from 7 to 10 May, and also the pressure for Europeans to federate being strongly urged at the time in the United States, notably by Senators Fulbright and Dewey, but he advised that the British Government "should seek to avoid either being led into foolish and premature expedients on the one hand and, on the other, a setback to the conception of West Europe which would give a stimulus to the Communist Party in France."

In defence of this attitude it can fairly be said that to support a policy of federation at a time when most Europeans would still not accept Germany as a partner on any terms, let alone on equal terms, was bound to seem to many — and particularly to experienced diplomats accustomed to view policies patiently in the long term — to be a "foolish and premature" expedient. Nevertheless it is hard not to feel that Kirkpatrick's ambivalent advice contributed to the fact that a great opportunity to advance the European movement and adopt federation as an ultimate goal was missed and that the British Government lost at that stage the leading role in the European movement which, by the successful efforts to create Western Union, it could fairly claim was Britain's up to the spring of 1948.

NORWAY'S "ATLANTIC POLICY"
The Genesis of North Atlantic Defence Cooperation

Olav Riste

After the publication of the relevant volumes of *Foreign Relations of the United States*, and with the opening of British public archives for the years 1948 and 1949, much interest has been focused on the diplomatic pre-history of NATO. Through contributions from historians as well as from some of the actors from that time, it has become clear that the first initiatives in the process that led to the North Atlantic Treaty date back to December 1947. Moreover, the prime mover is now seen to have been the British Foreign Secretary Ernest Bevin, with the Americans playing a more reactive and at first reluctant role.

What has remained generally unknown is that the idea of North Atlantic defence co-operation antedates Bevin's December 1947 initiative by almost exactly seven years. The origin of the suggestion that the nations bordering on the North Atlantic had vital defence interests in common and therefore ought to act together in peacetime for the protection of those interests lies in ideas put forward by Trygve Lie, Foreign Minister to the Norwegian Government-in-exile in London, to representatives of the British Foreign Office at the end of 1940, and later developed into Norway's wartime "Atlantic policy".

In December 1940 barely eight months had passed since the German occupation of Norway, an event which both the Norwegians and the Western Powers had been unable to prevent. Naturally, therefore, the direct motivation behind Trygve Lie's idea was to prevent a future repetition of that catastrophe. Yet the more timeless aspect of the proposal — the idea that nations separated by vast oceans had closely connected security interests requiring mutual protection — was a radical innovation. And it was this aspect which quickly became the centrepiece of Norwegian wartime policy planning.*

*For a fuller treatment, see the author's *London-regjeringa: Norge i krigsalliansen 1940–45* (2 vols, Oslo 1973–9).

Lie's first public intimation of Norway's long-term Atlantic security interests was made in a broadcast speech to Norway on 15 December 1940. Signalling a complete breach with Norway's non-aligned past, he stressed the need for post-war co-operation with nations sharing the same ideals and inspirations — primarily the United Kingdom and the United States — in order to safeguard the security of the Western world.

The Foreign Minister's speech was deliberately vague about forms of co-operation: speaking of the war-time alliance "which our allies and all progressive forces of the world are endeavouring to build up and strengthen", he went on to say that this endeavour would in its turn "provide the basis for a co-operation which can and must endure after the war: a political co-operation to secure our national freedom and remove the danger of assaults by arrogant and tyrannical aggressors, and an economic co-operation providing social security and preventing the destruction of our economies and our welfare."

Only between the lines can one here discern the far-reaching proposals for post-war military cooperation which the Foreign Minister was at the same time putting forward in private conversations with officials of the British Foreign Office: proposals pointing towards an alliance for mutual security covering the North Atlantic and embracing Britain, Norway, the United States, Iceland and the Faeroes. For the post-war world, Lie here envisaged nothing less than a joint defence arrangement with naval and military bases in the territories concerned.

Although post-war problems were hardly uppermost in British minds in those desperate times, the immediate reactions among British Foreign Office officials were distinctly favourable to Lie's ideas. They naturally welcomed the implied rejection of neutrality and isolation as a viable policy for the smaller European states in the future. And some saw the idea of a network of bases as being of particular interest. Thus Assistant Under-Secretary Sir Orme Sargent was inspired to put on paper the following long-term perspective:

One of the major post-war problems will be to enable this country to maintan its position vis-à-vis the Continent of Europe, and it is already fairly evident that the failure of France will render the co-operation of the United States essential for this purpose. May not the extension of M. Lie's ideas offer a practical means of achieving this co-operation? Just as there would be, according to M. Lie's suggestion, British and Norwegian [sic] bases in Norway, there might be similar Anglo-American bases in Portugal, Iceland and even Dakar, and at the same time American bases in Irish ports and even in British ports.*

Sargent was not alone in realising that the present salvation and future security of Western Europe hinged on the active engagement of the power of the United States. As seen by Lie's adviser and the chief architect of his policies, Professor Arne Ording, a central concern behind the "Atlantic policy" was in fact to attempt "to nail the Anglo-Saxon Great Powers to their responsibilities in Europe". For the time being, American neutrality of course barred any active exploration of the ideas with the leaders in Washington. Lie nevertheless in April 1941 took the opportunity of a talk with the US Minister to the Norwegian Government to acquaint a representative of the American Government with his thinking. Drawing a parallel between Norway's current importance to British maritime defence and her future importance to American strategic interests, the message he sought to convey was that in the age of fast, long-range aircraft the United States could no longer afford to be indifferent to the question of who controlled the Norwegian coastline. Hence his proposal that "the nations bordering on the North Atlantic should seek an arrangement or a plan for the future security and protection of the North Atlantic area."

Two months later, Lie had occasion to explain his views to a House of Commons audience at Westminster. Referring to his proposed security arrangement as "an Atlantic Association", he emphasized the need to have control of the seas as a key motive. But he also projected such a grouping as a possible nucleus for a collective security scheme which, by starting with "the nations that have not only fought together, but also belong together because they have not only common interests but also a common ideal in life and politics", would have a more realistic basis than the pre-war League of Nations.

Encouraged by the favourable albeit unofficial reaction of Foreign Office officials to his ideas, Lie in the autumn of 1941 took his proposals a major step further. In a lecture to a Royal

*Public Record Office, FO 371/29421, N 1307/87/30, 8 April 1941.

Institute of International Affairs audience at Oxford, later reworked into an article in *The Times* after consultations with Cabinet colleagues and others, Lie came closer to identifying some central aspects of the post-war international order which Norway would like to see established. He stressed at the outset, as a lesson of Norway's history as a maritime nation, "that the sea does not divide but links together. . . We are an Atlantic nation, and we do want, above all, a strong, organised collaboration between the two great Atlantic Powers, the British Empire and the United States of America."*

More specifically, Lie looked forward to an agreement on Eastern Atlantic defence with Britain, the United States and Canada, also covering Greenland and Iceland. He was convinced that Norway, after the war, would wish to be an active participant in such an arrangement — a conviction not shared by all his countrymen. Already Lie had confessed to British officials that some of his Cabinet colleagues as well as other influential Norwegians lagged behind in their understanding of the realities of international politics.

Among Lie's British interlocutors, too, the positive reactions to the "Atlantic policy" were tempered by scepticism as to its practicability. But the cautious official attitude of the British had much to do with the uncertainties surrounding the war aims of their new Great Power ally, the Soviet Union. Senior officials warned that Lie's "Atlantic policy" was "precisely the sort of post-war planning which Stalin is so anxious to discuss with us", and if he were to hear "that we, and possibly also the United States, were discussing it with the Norwegians without his being informed there would be the devil to pay."†

There was also concern that Stalin's post-war aims might include ice-free ports in North Norway. British Foreign Secretary Anthony Eden therefore advised Lie to proceed with caution. However, during Eden's talks with Stalin late in December 1941, no such Russian claims were put forward. On the contrary, Stalin's "spheres of influence" thinking appeared to envisage both a British-centred military alliance in north-

---

*The Times, 14 November 1941.

†Sir Orme Sargent in his minute of 14 November 1941 (Public Record Office FO 371/29422, N 6510/87/30).

western Europe and British naval bases in Norway and Denmark, as a counterpart to territorial adjustments and security arrangements for the Soviet Union along its western frontiers.

Encouraged by this, the British Foreign Office in the winter of 1942 moved closer towards approval in principle of a post-war security arrangement based on a system of military and naval bases as being "one of the few ideas of a post-war order which seem to have practical value and which have a chance of gaining general acceptance."* But the British Cabinet was still not prepared to take a stand on post-war security schemes in the absence of anything like a consensus among the three Great Powers which were now leading the Alliance. And Roosevelt, in particular, seemed averse to any "spheres of influence" arrangements.

In the meantime Lie and his advisers had continued to propagate their ideas, with markedly positive response from the Foreign Ministers of the Netherlands and Belgian Governments-in-exile. And by May 1942 the internal Norwegian debate on the proposals reached the stage where a full Cabinet consensus seemed within reach. After some concessions to both "universalists" and to advocates of closer co-operation at the regional Nordic level, the resulting official document, entitled "Principal Features of Norwegian Foreign Policy", showed that the Cabinet fully endorsed the "Atlantic policy" as the central feature of Norway's long-term security aims:

Until it becomes possible to create an effective and universal League of Nations, Norway will be compelled to seek security in regional arrangements. Norway, therefore, desires binding and obligatory military agreements concerning the defence of the North Atlantic, and she is anxious that Sweden should be a party to these agreements. The Norwegian Government would also look with satisfaction upon the adherence of Denmark, the Netherlands, Belgium and France to the system. The Norwegian Government desires that the military co-operation shall be developed as far as possible in the course of the war itself. The Norwegian Government desires to initiate negotiations even now regarding this future military co-operation.

It should again be stressed that the central motive for Norway's "Atlantic policy" so far was not to create a bulwark against

---

*Public Record Office, FO 371/32832, N 158/463/30, 26 January 1942.

Soviet expansion, but rather to take Norway out of its pre-war isolation and prevent a repetition of Germany's aggressive policies. By May 1942, however, and for a variety of reasons, no policy statement would have been complete without consideration of the relationship with the Soviet Union — and the Norwegian Government went on record with a strong endorsement of closer relations and co-operation with the USSR. Speculations about Soviet designs on North Norway were denounced as groundless, and the official policy document seemed to recognise a legitimate Soviet security interest in the far north:

During this war, Northern Norway has been one of the starting points for the German attack upon the Soviet Union. Provided that there is friendly co-operation between the Soviet Union and the Western Powers, the Soviet Government will be positively interested in the development of the defence of Northern Norway. Should the relations between the Soviet Union and the Western Powers be hostile, the position of Northern Norway would be much more complicated. Norway will, therefore, do her utmost to prevent such a conflict from arising.

In sum, this document made clear both Norway's post-war security aims and her basic dilemma at the crossroads of Western and Soviet strategic interests. The dilemma was more succinctly recorded in the diary of Lie's adviser, Professor Arne Ording, later that year, when he speculated about the ramification of the "bases" idea: "We may find ourselves faced with the following choice: either to provide bases only for the British and the Americans, which the Russians may see as a threat, or to give them also to the Russians, which will create both strategic and internal problems. Or to adopt a new neutrality line, this time balancing between the Western Powers and Russia in the same way as we previously, with tragic and well-known results, tried to balance between Germany and the Western Powers."

In many ways this document from May 1942 represented a climax in the development of Norway's "Atlantic policy". From then on, as the post-war planning of Britain and the other Great Powers got into its stride, Norway's role as the initiator of new policy approaches was transformed. Henceforth, the prime necessity became that of adjusting to policy developments determined by the Great Powers.

To some extent, the Norwegian Government also had to face the problem of internal dissenting opinions, as well as a revival of interest in closer Nordic co-operation. In liberal and conservative Norwegian circles outside the Government, the isolationist impulse was not yet dead. From the United States, the Norwegian conservative leader C.J. Hambro attempted to re-establish the old distinction between the cynicism of the Great Powers and the moral superiority of small states, and warned against a peace settlement dictated by the "Big Four".

On the Nordic side, Swedish advocates of a Nordic federation during 1942 and 1943 found the time opportune for renewed calls in favour of Nordic unity after the war. As the Norwegian over-reaction against the idea showed, the attempt could hardly have come at a worse time. For just as the Swedes could only discuss Nordic unity on the basis of complete neutrality as between the Great Powers, the Norwegians could only discuss it within the framework of co-operation with the Allies. For either of them to discuss Nordic unity on any other basis would mean sowing the seeds of doubt about the firmness of their current policies.

In the autumn of 1942, Anthony Eden tried to extract from the British Government some sort of commitment to an Atlantic defence system. Both the Netherlands and Belgian Foreign Ministers were now on record as supporters of the idea of tying Great Britain and the United States into a Western post-war security system. And the Dutch Foreign Minister Van Kleffens, after a visit to the United States, claimed to have found much positive interest in the idea among American leaders.

Prompted by this widening support, Eden, in October 1942, attempted to get the agreement of the War Cabinet to instructions to the British Ambassador in Washington to discuss the idea with the State Department and to express general British support for the political content of this projected defence system. But the consensus of the War Cabinet was opposed to even such a limited commitment on the part of Britain, and the outcome was a request that the Foreign Office first develop the broader lines of British and international post-war security.

The resulting memorandum, entitled "The Four-Power Plan", gave primary emphasis to the emerging American ideas of a universal system of post-war security and co-operation dominated by the Great Powers. Regional defence systems had their place: "In particular, it is to be hoped that in North Western Europe special agreements will be made whereby it will be possible for Great Britain and possibly the United States to establish naval and air bases in the territories of the various powers bordering on the North Sea. . ." But regionalism seemed by now to have been relegated to a somewhat ancillary concept.

These new signals were also received in the Norwegian Foreign Ministry. In November 1942 Professor Ording noted in his diary that "we must now put the main emphasis on the United Nations." And in March 1943, during Trygve Lie's first visit to Washington, Sumner Welles, then Under-Secretary of State, told him that regional defence arrangements would have to await the erection of a global security system. "One had to start with the greater framework in order to avoid the controversies that could easily arise if one started at the other end." Welles also led Lie to understand that the United States and Britain would primarily be interested in links with a unified Nordic group.

Similar conclusions could be drawn from Winston Churchill's first public statement on the post-war international order. In a radio speech on 21 March 1943 he projected a vision of groupings of small states with strong links to the major Powers on whom would rest the overall responsibility for the maintenance of peace. However, while the Norwegian Foreign Ministry clearly saw the need for an alteration of priorities in its foreign policy, the uncertain contours of the Great Powers' plans for the post-war world during most of 1943 precluded any major restatement of the main policy lines.

In January 1944, the change of emphasis in Norwegian foreign policy for the post-war world was publicly stated both in a major speech by the Foreign Minister and in a parallel article in *The Observer* by Professor Ording. First priority was now clearly given to the universalist concept of the United Nations. And this was done, as Lie admitted to the Foreign Office, in order to conform to the views of Great Britain and the United States. An Atlantic regional arrangement was still stated as most likely to provide the maximum degree of security for Norway in the post-war world. But this took second place to the global scheme of four-power co-operation, and was, moreover, made dependent on the goodwill, or at least lack of opposition, of the Soviet Union. As Lie put it, "Norway's interests would be best served by an agreement embracing the countries bordering on the North Atlantic, on condition that it was subordinated to an international organisation and was accompanied by an extension of our good relations with the Soviet Union."

This, then, was the general order of priority applied by the Norwegian Government in its approach to the organisation of peace during the remainder of the war. But the new proviso regarding relations with the Soviet Union points to the final and major development in Norwegian war-time foreign policy: the emerging "special relationship" between Norway and its new Great Power neighbour to the east.

In fact, ever since the autumn of 1942 the Soviet Union had been showing an increasing interest in Norwegian foreign policy guidelines. And in April 1943, while enquiring of the Norwegian Finance Minister whether there were any new developments in Norway's "Atlantic policy", the Soviet Ambassador to the Norwegian Government-in-exile said that "Norway should be aware that in order to obtain their security aims, it was not only necessary to be in agreement with the Western Powers. One should, in addition, make sure of a good relationship with the Soviet Union, which was also a Power with Atlantic interests."

But the major impetus for an increasing attentiveness to Soviet interests in the north was provided by the prospect that the Red Army, in the eventuality of Finland withdrawing from the war, might become the first Allied liberation troops on Norwegian territory. This prospect, and the lack of interest on the part of the Western Powers in providing Allied liberation forces to counter-balance the possible presence of Soviet troops in North Norway, lay behind a series of co-operative approaches towards the Soviet Union simultaneously with similar agreements with the Western Powers — agreements intended to regulate the exercise of Allied military jurisdiction on Norwegian soil and the gradual transfer of sovereignty to Norwegian authorities.

Soviet policy during this period showed a clear preference for conducting Soviet-Norwegian relations in regard to North

Norway on a strictly bilateral basis, excluding any involvement of the Western Great Powers. However, the possibility of establishing bilateral relations and closer co-operation on a more permanent basis was wrecked when Molotov, during a midnight meeting with Trygve Lie in the Kremlin in November 1944, presented his demands for a revision of the Svalbard Treaty and the cession of Bear Island to the Soviet Union. These demands inevitably produced shock waves that were to reverberate far into the post-war period.

The immediate effect of Molotov's brusque tactics was to pulverise the prospects of a cordial relationship being established in the north. The presence of Soviet troops on Norwegian territory in Finnmark, initially welcomed by the Norwegian Government in an official statement on 26 October as "a further manifestation of the friendship between our two countries", was henceforth bound to serve as a reminder that Soviet aims and Norwegian territorial sovereignty in the Arctic might be at cross-purposes.

The result of these developments was that the Norwegian Government, in the course of 1944, passed from an early restatement of its Atlantic policy preferences, through a period of accommodation to Soviet interests, to an attitude of detachment from international politics. From the autumn of 1944 until the end of the war, the prevailing mood in the Norwegian Government was one of disillusionment with the Western Powers, and deep suspicion of the aims of Soviet policy in the north.

With such a fundamental uncertainty about the future direction of Norwegian foreign policy, it is hardly surprising that the belated British attempts during 1944 to regenerate interest in Atlantic security schemes should fall on stony ground as far as Norway was concerned. The time for Norway's open commitment to collective Western defence arrangements had passed. Instead, Norway seemed on the way to fulfilling the pessimism of a Foreign Office instruction to the British Ambassadors to Norway and Sweden in June 1943. Explaining why the British Government was reluctant to take sides in the persistent disputes over the merits or otherwise of Nordic union after the war, this instruction went on to say: "We must not forget that Norway was an ardent neutral until she was invaded by Germany, and that she is a belligerent in spite of herself. There is no reason to suppose that she will not resume the mentality of neutralism as soon as the war is over, notwithstanding the preachings of Monsieur Lie to the contrary."

Norway's return to formal nonalignment in 1945, however, did not signify a return to the *status quo*. The basic premise of Norway's "Atlantic policy" — a realisation that neither large nor small European nations would in the future be able to opt out of the international power game — remained in force, despite a recurring nostalgia for times past when Norway had seemed able to keep out of other nations' quarrels. Moreover, the disillusionment towards the Western Powers may have concealed, but could not displace, a strong counter-current of practical and functional ties developing from the war-time association, whether in the form of arrangements concerning supplies and training for Norwegian forces or through Norway's participation in the Allied occupation of Germany. Such arrangements, while devoid of a formalised security policy superstructure, preserved the central element of the nation's security "lesson" from the war: the conviction that in order to restore Norway's independence in case of assault by a major power, allies would be needed, and that military co-operation could not be left to improvisation.

THE OMAHA MILKMAN
The role of the United States in the negotiations
Theodore C. Achilles

"I am convinced that the Soviet Union will not deal with the West on any reasonable terms in the foreseeable future and that the salvation of the West depends upon the formation of some form of union, formal or informal in character, in Western Europe, backed by the United States and the Dominions — such a mobilisation of moral and material force as will inspire confidence and energy within, and respect elsewhere." (*Ernest Bevin to George Marshall, 15 December 1947.*)

The above words were spoken privately by the British Foreign Secretary to the American Secretary of State the evening the two-year attempt of the Council of Foreign Ministers to draft peace treaties with the Axis powers finally broke down due to Soviet intransigence. Bevin expressed similar views a few days later to Georges Bidault, then Foreign Minister of France, and publicly in the House of Commons on 22 January 1948.

In December 1947 much of Western Europe was still prostrate materially and economically from the war. Allied demobilisation had been swift and far-reaching. The Soviet army was on the Elbe, the formerly free countries of Eastern Europe were occupied and subdued, and the Soviet shadow lay darkly over Western Europe. Truly there was need to "inspire confidence and energy within, and respect elsewhere".

As a step toward Bevin's objective, Belgium and the Netherlands were invited in January 1948 to adhere to the Treaty of Dunkirk, which already linked Britain and France in an alliance against renewed German aggression.* Paul-Henri Spaak, the Belgian Foreign Minister, replied that such a pact would be meaningless unless it were designed for protection against Russia and unless the US were included. When Spaak so advised us in Washington, we suggested he read the Treaty of

---

*France and the United Kingdom had signed a 50-year Treaty of Alliance and Mutual Assistance at Dunkirk on 4 March 1947.

Rio de Janeiro. (The Rio Treaty was a "collective defence agreement" under the UN Charter linking the 21 American republics. Senator Vandenberg, Chairman of the Foreign Relations Committee had been involved in negotiating it and the Senate had recently approved US ratification.)

The views Bevin had expressed privately to Marshall were formally communicated to the US in a memorandum from the British Ambassador, Lord Inverchapel, on 13 January. John D. Hickerson, Director of the State Department's Office of European Affairs, advised Marshall to reply to Bevin that he considered the latter's objective "magnificent" but extension of the Dunkirk Treaty against Germany "a dubious means". He recommended that Bevin be urged instead to seek a European Treaty modelled on the Rio Treaty and to say that, if this were negotiated, the US might adhere.

Marshall was hesitant. In a brief note handed to the British Ambassador he merely expressed a "hearty welcome" for this European initiative. He did, however, authorise Hickerson to convey the latter's views, as representing "those of the Bureau of European Affairs", informally to the Ambassador.

There was ample reason for caution. George Washington's warning in his farewell address against "entangling alliances" was taught to every American schoolchild and often reiterated in political oratory. 1948 was an election year. There was a Democratic President and a Republican majority in Congress. Senator Vandenberg was himself considered by many to be a contender for the Republican nomination to run against President Truman. Congress had not yet approved the legislation and appropriations necessary to give effect to the Marshall Plan for European reconstruction.

In the circumstances, the US stance toward Bevin and Bidault's insistent inquiries as to US support was, for some months, in effect: "Show us what you are prepared to do for yourselves and each other and then we will see what we can do."

Hickerson's advice, given informally to Lord Inverchapel as the "views of the Bureau of European Affairs", was nevertheless accepted by Bevin and pressed by him upon the French and Benelux Governments. The Brussels Treaty was signed on 17 March and was, like the Rio Treaty which served as its model, a collective defence agreement within the UN Charter.

The day the Brussels Treaty was signed, President Truman addressed the Congress, calling the Treaty one "of great significance" deserving "the support which the situation requires", and stating, "I am sure that the determination of the free countries of Europe to protect themselves will be matched by an equal determination on our part to help them to do so."

Nevertheless, in response to British and French pleas for concrete evidence of US support, the US position continued to be "show more effectively what you can do for yourselves and each other," specifically by implementing the Brussels Treaty. Accordingly, Western Union military headquarters was established at Fontainebleau with Field-Marshal Montgomery as Supreme Commander. It was a step, but the Supreme Commander had little to command. British Embassy friends told us of a message from him to London saying, in effect: "My instructions are to hold the line of the Rhine. With present forces I might be able to hold the tip of the Brittany Peninsula for 48 hours. Please instruct further."

There was no disagreement as to the need for military as well as economic assistance from the United States. Representatives of the Joint Chiefs of Staff were promptly sent to Fontainebleau to work with Montgomery and his planners as "non-participating members".

Beyond this, opinion in the US Government was naturally divided with respect to such a drastic change in policy as entering into a military alliance in peacetime. Strong voices in both the State and Defense Departments opposed the idea vigorously; public opinion would never support it; the Senate would never approve ratification; military and economic assistance would be enough without the formal ties involved in a treaty relationship. Others opposed US "participation" on an equal footing, preferring only some sort of bilateral agreement between the US and Canada, as one unit, and the Brussels Treaty nations as the other. (The last became known as the "dumb-bell theory" in reference both to its shape and its intellectual content.)

Hickerson led the fight within the US Government for full participation. He also insisted on the importance of keeping the process of working toward an alliance fully bipartisan and of maintaining consultation with the Senate Foreign Relations Committee during the negotiations. Both stipulations proved invaluable.

Meanwhile, recognition of the importance of full US participation was growing. The Soviet seizure of Czechoslovakia, the threat to Norway, and the Communist efforts to disrupt the economic reconstruction of Europe were affecting opinion on both sides of the Atlantic. In April, the National Security Council approved a State Department recommendation that the President announce, after diplomatic soundings, that the US was prepared to negotiate a collective defence agreement both with the Parties to the Brussels Treaty and with Norway, Denmark, Sweden, Iceland and Italy, and that, pending conclusion of such an agreement, it would regard an armed attack against any member of the Brussels Treaty as an armed attack against the United States.

Senator Vandenberg's reaction was sought. John Foster Dulles, as Vandenberg's representative, had been on the US delegation at the Council of Foreign Ministers' meeting in London, had accepted Bevin's view as quoted above, and had convinced Vandenberg of its wisdom. Vandenberg favoured an alliance, but objected strongly to a presidential initiative in this connection.

"Why should a Democratic President get all the kudos in an election year?" he asked. "Wouldn't the chances of Senate 'consent' to ratification of such a treaty be greatly increased by Senatorial 'advice' to the President to negotiate it?" Vandenberg was not only partisan, he was truly a statesman.

In 1948 American opinion toward the United Nations was a mixture of hopes for its success and recognition of its weakness. A number of proposals for strengthening it were pending in Congress. To capitalise on this sentiment, Senator Vandenberg, with our assistance, drafted the "Vandenberg Resolution", which, among several proposals for strengthening the UN, included three key paragraphs:

Progressive development of regional and other collective arrangements for individual and collective self-defense in accordance with the purposes, principles and provisions of the Charter.

Association of the United States, by constitutional process, with such regional and other collective arrangements as are based on continuous and effective self-help and mutual aid, and as affect its national security.

Contributing to the maintenance of peace by making clear its determination to exercise the right of individual or collective self-defense under Article 51 should any armed attack occur affecting its national security.

The Resolution was adopted by the Senate on 11 June 1948 by an overwhelming vote. The Senate thereby "advised" the Administration to negotiate an alliance which would, by its nature, *deter* aggression.

Still we proceeded cautiously. The negotiations continued for months and were largely completed before we were prepared to admit publicly that they were in progress. Early in July, the "Washington Exploratory Talks on Security" began. The first meeting was convened by Acting Secretary of State Robert Lovett with the British, French, Netherlands, Canadian, and Belgian Ambassadors, the last-named also representing Luxembourg, each accompanied by two or three assistants. It was these assistants who constituted the "Working Group" which actually negotiated the Treaty. Meetings of the Ambassadors with the Acting Secretary were held occasionally to review the progress of the Working Group, particularly when it was necessary to get an obstinate government to modify its position.

The process was deliberately a leisurely one since the US team made clear the importance it attached to avoiding public controversy until after the presidential election in early November. In September, the Group submitted to the respective Governments a memorandum tentatively recommending conclusion of a collective defence arrangement, discussing its potential membership and geographic coverage, and appending as an annex an "Outline of provisions which might be suitable for inclusion" — in effect, a draft treaty with certain blanks and alternative paragraphs.

At the end of October, the Brussels Treaty Governments formally stated their intention to participate in negotiation of such a treaty, and the negotiations continued until it was ready for signature on 4 April 1949. The difference was that after the elections in early November, Hickerson and I met regularly with the Senate Foreign Relations Committee to discuss actual treaty language, and even more frequently with the Committee's Chief of Staff, Dr Francis O. Wilcox. By the time the Treaty was completed, the Senators had acquired a vested interest in it.

The negotiations throughout were informal in the extreme. They were conducted by a small group, at the working level, sitting around a small table, in a Washington summer. It was "shirt-sleeve diplomacy" literally and figuratively. There was a strong common desire to reach agreement and to find the best mutually acceptable answers. Derick Hoyer Millar, then Minister in the British Embassy and now Lord Inchyra, started what became known as "the NATO spirit". One day, he made a proposal which the rest of us criticised severely but constructively. Derick replied: "Very well. Those were my instructions. I'll tell the Foreign Office I've made my pitch and been shot down, and ask for instructions along the lines we've agreed."

Hickerson made another contribution. He maintained that the treaty should be written in such simple language that "even a milkman in Omaha can understand it." The "Omaha milkman" unwittingly became the stylistic godfather of the Treaty.

There was no disagreement as to the general nature of the new treaty, its relationship to the UN Charter, or that it was to be modelled on the Rio and Brussels Treaties. There was naturally a great deal of argument over various specific provisions, particularly those with respect to commitments, geographic coverage and duration, and which other governments should be invited to become parties.

The basic differences were due to the fact that the Europeans, particularly the French, wanted as binding and as long a commitment as possible, and the Americans, while agreeing in principle, were constrained by what the Administration and friendly Senators thought the Senate would accept.

Article 5, specifying the commitment to respond to armed attack, was naturally the most controversial. On this point the Rio Treaty had provided: ". . . an armed attack . . . against an American State shall be considered an attack against all the American States and, consequently, each . . . undertakes to assist in meeting the attack."

In Article IV of the Brussels Treaty, the language was that the Parties would ". . . afford the Party so attacked all the military and other aid and assistance in their power."

It took many sessions of the Working Group, much consultation with Foreign Offices (and with the Senate Foreign Relations Committee) to reach the formula in Article 5 of the North Atlantic Treaty that ". . . an armed attack against one or more . . . shall be considered an attack against them all and . . . that each . . . will assist the Party or Parties so attacked by taking forthwith, individually and in concert with the other Parties, such action as it deems necessary, including the use of armed force, to restore and maintain the security of the North Atlantic area."

The words "such action as it deems necessary" were insisted upon by the Senators because the US Constitution provides that only the Congress may declare war. The words "forthwith" and "including the use of armed force" were intended by the Americans, and accepted by the Senators, as a binding commitment upon the Executive to take immediate action, including military action, with the belief that the Congress could be counted upon to back it up with a declaration of war if the armed attack were not just an incident but a fullfledged act of war.

The critical element in the negotiation of the North Atlantic Treaty was therefore the formula defining the commitment of each Party to act individually or collectively in the event of an armed attack against any other Party — a formula which would be acceptable both to the US Senate and to the participating European Governments.

When that formula, embodied in the final text of Article 5, had been agreed, the negotiators took up the problem of *where* an armed attack would bring the obligations of Article 5 into play. The Treaty area must obviously cover the territory in Europe and North America of the Governments initially involved in the negotiations — Britain, France, Belgium, the Netherlands, Luxembourg, Canada and the US — and the sea and air spaces between them.

The geographic area to be covered by the Treaty naturally depended upon which countries would become Parties. But what other nations would be desirable allies? Which would strengthen the common security? What of overseas possessions?

There had already been approaches from other governments indicating a desire to participate, and many bilateral discussions with them. It had been the consensus in the negotiating group from the beginning that Norway, Denmark, Sweden and Iceland would be desirable and helpful participants. We knew there was little chance of Sweden abandoning its neutrality and we worried lest Sweden might succeed in enticing the other Nordic countries into what we considered an inadequate Nordic defence agreement.

US insistence on the importance of Norwegian, Danish and Icelandic membership was due not only to the depth of their democratic heritage and the necessity of safeguarding the Northern flank of Europe but also to the "stepping stones" concept. In those days of comparatively short-range aircraft it was considered that US reinforcement of its European allies in the event of war would require refuelling facilities in Greenland (i.e. Danish adherence), Iceland and the Azores. This indicated the importance of Portuguese membership and Portugal was duly invited. Portugal was at first reluctant, suspecting that its involvement might conceal British and French designs on its colonies, but accepted our argument that the Treaty represented an "Atlantic" rather than a "European" concept. Ireland was also invited, but when she replied that she would be glad to join if Britain returned Ulster to the Free State the matter was dropped.

Italian membership was more controversial and was opposed by various governments, and in some quarters in the US, as going beyond the strictly Atlantic framework and thereby opening the gates for other nations to apply. It was recognised that Germany in particular and Greece, Turkey and Spain were potential future members.

The Tropic of Cancer was adopted as the southern boundary of the treaty area simply to avoid involving any part of Africa or any other of the American Republics (Mexico, Cuba or any others). It is worth recalling today that the Treaty area, as defined in Article 6, is simply that in which an armed attack would constitute a *casus belli: there was never the slightest thought in the mind of the drafters that it should prevent collective planning, manoeuvres or operations south of the Tropic of Cancer in the Atlantic Ocean, or in any other area important to the security of the Parties.**

The British, French, Dutch and Belgians would have liked

---

*Author's emphasis.

some commitment for assistance in the event of attack on their overseas possessions but realised that this would have aroused insuperable opposition in the US Senate. Agreement was easily reached on Article 4, that "the Parties will consult together whenever, in the opinion of any of them, the territorial integrity, political independence or security of any of the Parties is threatened." It was understood by all that the scope of this Article was worldwide.

The Canadians in particular, and some of us on the US side, strongly believed in the truth of a statement by Senator Vandenberg: "Unless the Treaty becomes far more than a purely military alliance it will be at the mercy of the first plausible Soviet peace offensive."

The Europeans were not enthusiastic about provisions for cooperation in non-military fields except for some French interest in cultural cooperation. Yet the Annex to the Working Group's memorandum of 9 September recommended "provision for the encouragement of efforts between any or all of the parties to promote the general welfare". When this language was discussed informally with the ranking members of the Senate Foreign Relations Committee they objected strongly on the grounds that the words "general welfare" had caused more litigation than any other words in the US Constitution. The present language of Article 2 was then drafted by the Canadians and ourselves and accepted by the Europeans. NATO has never yet developed this article, except in a minor way through the Committee on the Challenges of Modern Society, but it is still, as it was intended to be, a reservoir of great potential.*

---

*Despite its interest in the inclusion of such an article in the Treaty, the Canadian Government once passed up a potentially far-reaching opportunity to give it effect. Shortly before the Ministerial Meeting of the Council in Lisbon in February 1952, the United States had imposed a minor restriction on the import of woodpulp, which was of concern to Canada. At the Lisbon meeting Lester Pearson, then Canadian Minister of External Affairs, asked me to ascertain informally what Washington's reaction would be if Canada were to invoke the obligation in Article 2 for the Parties to "seek to eliminate conflict in their economic policies". Quick soundings revealed that Washington would not mind in the least: on the contrary, it would welcome a friendly invocation of Article 2. Unfortunately Ottawa got cold feet and the opportunity was missed.

Of course, economic consultation has long been a part of the Alliance's

The duration of the Treaty was one of the last points to be agreed. The US, fearful of Senate rejection of anything more than a brief commitment, held out for a duration of ten years. The French insisted on a duration of fifty years and had some support from other Europeans. The final language, providing for unlimited duration but opportunity for revision after ten years and possible withdrawal after twenty, has stood the test of time.

During the negotiations, little importance was attached to a preamble — "mere pious platitudes" — but the thinking which inspired Article 2 also led to the words in the Preamble: "They are determined to safeguard the freedom, common heritage and civilization of their peoples, founded on the principles of democracy, individual liberty and the rule of law."

The fundamental common values of the Parties, tersely summarised in the Preamble, are probably more responsible than anything else for the enduring quality of the Alliance.

In retrospect, it may be worth noting a few points which were considered and deliberately omitted from the Treaty.

One was a provision for expulsion should a member be subverted by extreme right or left. It was omitted in the belief, perhaps unwarranted, that any government taken over by totalitarian elements would probably withdraw voluntarily.

Another was any provision for rules of procedure. This was avoided to ensure flexibility and particularly to *avoid any requirement for unanimity* or other fixed voting. The theory was — and it is as sound today as it was then — that *no government could be forced to take action against its will* but that, conversely, *no government could prevent other governments from taking such collective action as they agreed to take.* A third was any provision for an international staff. Wise Senators argued strongly against Parkinsonian development of a new international bureaucracy. The thought of the negotiators was that there should be no international staff. Rather, modest

---

activities. It led, for example, to Council agreement in December 1978 on the urgent necessity of increasing financial assistance to member countries facing particular economic difficulties, an initiative which was followed by the adoption under the auspices of the OECD — a forum more suited for such action — of a $1.5 billion assistance programme for Turkey. Nearly all of this sum was pledged by NATO member-countries.

national delegations would be housed in the same building and together do whatever was necessary. Alas for our innocence! It worked for a few weeks. Then it became clear that we needed an accountant to pay the bills, then we needed a press officer, and then Professor Parkinson moved in.

As far as the United States was concerned, the process followed in concluding the Treaty was an example of bipartisanship at its best and of close cooperation between the Executive Branch and the Senate. The Treaty, involving so radical a change in US relations with Europe that the sceptics at first gave it minimal chance of Senate approval, was ratified after a vote of 82 to 13.

On 4 April 1979, the United States Congress unanimously adopted a Resolution reading in part: "*Resolved by the Senate and House of Representatives of the United States of America in Congress assembled,* that the North Atlantic Alliance be reaffirmed as a vital commitment and cornerstone of United States foreign policy, and that the bipartisan spirit that inspired its birth be rededicated to the purpose of strengthening it further in the cause of peace and security."

Also marking the 30th Anniversary, a Proclamation by President Carter declared in part:

"Since NATO's inception, the international situation has evolved in many respects and NATO has adapted to these changes — militarily, politically and economically. Today the Alliance remains as relevant and centrally important to our security and way of life and to the independence of the United States as it was in 1949. Then as now, the firm support of Congress and the American people for NATO reflects their deep conviction that NATO is the cornerstone of United States foreign policy.

"As NATO moves forward into another decade of achievement, we look toward the future with confidence, aware that continuing Allied cooperation will provide the international stability and security upon which our ideals, our civilization, and our wellbeing depend. As NATO begins this new chapter in its distinguished history, I am proud to rededicate the United States to the NATO objectives which have served the cause of peace so well."

In retrospect, I believe that those of us who negotiated the Treaty did well, partly in spite of ourselves. We did well, deliberately, in formulating a simple, flexible instrument designed to provide the basis for the development of as much military, political and economic unity as the respective governments may be prepared to accept at any given time. What we recognised only dimly was the strength, depth and lasting importance of our common heritage, interests and basic human values.

SEARCHING FOR SECURITY IN THE NORTH

Denmark's road to NATO

Sven Henningsen

The years 1948-9 were ones of shattered illusions in Danish foreign policy and the beginning of a new epoch in Denmark's place in the international world. "No new 9th of April" was a generally held opinion of the Danish people in May 1945 at the end of World War II. This alluded to the German occupation of the country on that date in 1940, followed by the "five evil years", and expressed a desire to break with decades of isolated neutrality.

The defeat of Germany is 1945, and the extension of Soviet power over a great part of the eastern and southern shores of the Baltic, radically changed Denmark's security problems. In his conversations with Hitler in November 1940, the Soviet Foreign Minister, Molotov, had made it quite clear that the Soviet Union would be interested in the entrance to the Baltic through the Danish Sound and Belts. This, however, did not influence Danish reflections on foreign policy issues in the immediate post-war period. The main problems at that time were the withdrawal of the Soviet troops which had liberated the island of Bornholm from German occupation during the last days of the war, the evacuation of approximately 300,000 civilian German refugees, and an impassioned discussion on the future status of the old Danish territory of South Schleswig, which had been conquered by Prussia in the war of 1864, but still had a minority of Danes among its population.

The Danes, in general, shared the illusion of the victors and the liberated nations that the cooperation and friendship between the victorious Great Powers would continue into the post-war period. These hopes were concentrated on the United Nations, which would guarantee peace and security for all members. The failure of the League of Nations in the 1930s had led to Denmark's isolated neutrality. Such a policy was rejected after 1945, but the outcome of conflicts between the Great Powers in the Security Council, and the ensuing stalemate, created disappointment, and the result was renewed small power neutrality in the growing East-West conflict. Leading Danish politicians proclaimed that Denmark would not join any power bloc but would follow a policy of "bridge-building" between East and West, and an active UN policy. This attitude found expression in the decision to postpone the rebuilding of the Danish armed forces until the Security Council had found a solution to the problem of a UN army.

Developments in Eastern Europe and the Soviet Union's policies in East Germany and other parts of Europe, with the outbreak of the Cold War, gradually destroyed the hopes of a harmonious post-war world, with the result that there was a change in Danish foreign policy from neutrality to participation in an alliance with like-minded countries.

Economic ties with the West and basic interest in cooperating in rebuilding the Western European economies made Danish participation in the Marshall Plan and the Organization for European Economic Co-operation (OEEC) almost self-evident. Denmark's considerable foreign trade depends on the greatest possible economic freedom and cooperation in the world, most of all between the Western industrialised nations. The changing climate in international politics shattered the illusions of a peaceful and secure future.

Traditionally Denmark belonged to the Western world and in a more narrow sense to Western Europe. The Danish people shared the political and cultural values of the peoples of Western Europe and North America. Danish socio-economic life was, as in several other Western European states, built on a belief in a "mixed" liberal-capitalist economy. The growing social concept of a welfare state also corresponded to similar tendencies elsewhere in Western Europe. Common experiences during World War II strengthened the ties to the West and the desire to protect free institutions and the right to independence and self-determination.

Increasingly, political events in 1948 seemed to endanger these common values and interests. In January, the Soviet press attacked Denmark for placing its armed forces under American and British control. The Communist coup in Czechoslovakia in February made a deep impression in Denmark as elsewhere in the West. Czechoslovakia had been considered a "bridge-builder" in an increasingly divided Europe, and events there seemed to prove that a democratic system could not survive within the Soviet-dominated sphere. Although the coup was administered by the Czechoslovak Communist Party, there was general agreement among Danes that it was inspired and directed by the Soviet Union. Within Denmark there was deep resentment against the small but vociferous Communist Party which defended the coup. Equally negative effects were to be produced by the Berlin blockade and the airlift which began in June that year.

The interest of the decision-makers in the Kremlin turned towards Scandinavia early in 1948. On 27 February President Paasikivi of Finland received a letter from Stalin proposing a treaty of friendship, co-operation and mutual help, similar to the Soviet treaties with Hungary and Romania. The Finnish answer was dilatory, but in Scandinavia the Soviet proposal was considered the first step towards Finland's complete absorption into the Soviet bloc. On 17 March President Truman declared in Washington that the Soviet Union's pressure on Finland involved a danger to all the Scandinavian states. There were rumours of a similar Soviet proposal to Norway.

In Denmark, as well as Norway and Sweden, the Soviet proposal to Finland added to the anxieties about the Soviet Union's intentions, and they were not mitigated by the treaty which the Finns signed on 6 April 1948. Fears in Denmark were further exacerbated by the so-called "Easter crisis". A report from the Danish Minister in Washington, Henrik Kauffmann, based on information from American sources, indicated that a Soviet threat was imminent against Western Europe, including Norway and Denmark. The reaction in Copenhagen was to increase military preparedness during Easter, but although the alarming report turned out to be false, it put the fundamental question of Denmark's future security on the agenda.

The Social Democratic Government with Hans Hedtoft as Prime Minister had various options: to continue the policy of neutrality — which was rejected; to seek a guarantee from the United States and the United Kingdom; to obtain membership in the Brussels Treaty alliance; or to adopt either an "Atlantic" solution or a Scandinavian defence arrangement. The preference of the Government and the majority of the Danes was for a Nordic solution.

Strong historical, cultural, ideological and political traditions bound the Scandinavian nations together. Until the beginning of the nineteenth century they had often fought wars with each other, but from the 1820s an active movement for a closer union began, born out of the Romantic movement and supported by students, the national liberal parties and the royal families.

The Scandinavian movement had strong foreign policy overtones. The Danes sought support against German nationalism in the duchies of Schleswig and Holstein, where German nationalists, supported from Germany, wanted to break away from the connection with Denmark and form an independent German state. The unsuccessful discussions about a Scandinavian alliance with the aim of a future union left Denmark alone to fight the war with Prussia and Austria in 1864, with the subsequent loss of the duchies. From then on, the Scandinavian movement sought more limited and practical results. During World War I, all Scandinavian countries remained neutral, cooperating economically. In the 1930s, regular meetings between the Foreign Ministers tried with small success to chart out a common foreign policy, and suggestions from some Danish and Swedish politicians and journalists to investigate the possibilities for a military alliance were rejected by the Governments.

During World War II, Denmark and Norway became, in April 1940, victims of Nazi aggression, while the Faroes and Iceland were occupied by the United Kingdom, and Greenland by the United States. Sweden remained neutral, but Finland fought the Soviet Union in the Winter War of 1939–40 and, from 1941 until 1944, was allied with Germany — an illustration of the conflicting strategic positions and security interests of the Scandinavian countries.

When the Norwegian King and his Government went into exile in London, the already strong Norwegian ties with the West were strengthened. Norwegians also took part in the informal discussions about a future form of Atlantic co-operation. In occupied, isolated Denmark the only link with the free world was Sweden. Danes listened to Swedish radio, and read Swedish newspapers. When the German grip on domestic affairs tightened, fleeing members of the resistance movement crossed the Sound and found asylum in Sweden, as did the majority of

Danish Jews in the autumn of 1943. In the final period of the occupation, the Danish resistance movement received military and economic support from the Swedish Government. Thus when people in occupied Denmark discussed post-war international problems, Scandinavian co-operation — in particular with Sweden — took a central place.

It was, however, the Norwegian Foreign Minister Halvard Lange who took the initiative that set off the negotiations in 1948. In a speech on 19 April, he gave expression to his interest in a Scandinavian defence community, but indicated simultaneously the possibility of Norwegian accession to a Western bloc without Sweden and Denmark. He said that "the strategic and security problems, which each of the Scandinavian states is confronted with, are not identical, and this fact may involve certain difficulties, in the case of finding a common solution." The effect in Sweden was a resolution in the Parliament's Foreign Relations Committee authorising the Government to initiate negotiations with Norway and Denmark about the possibilities of establishing such a Scandinavian defence association.

The Norwegian and Danish Governments accepted the invitation, and on 10 May the Prime Ministers met in Stockholm to make a preliminary survey of the problems involved. From the beginning of the discussion it became obvious that there were fundamental differences between the Swedish and Norwegian concepts. The Norwegian Government accepted the idea of a Scandinavian alliance if some form of understanding with the Western bloc could be obtained. If that were impossible, the Norwegian Foreign Minister felt that Norway's best interest would be to join an Atlantic pact. The Swedish view was that the purpose of a Scandinavian defence alliance was to preserve Scandinavia's freedom of action, or, to put it another way, to keep Scandinavia neutral between the Eastern and Western Great Power blocs. The Danish Prime Minister, Hans Hedtoft, who was not only politically but also emotionally committed to the Nordic movement, tried through all the negotiations to mediate between the Norwegian and Swedish points of view.

The Prime Ministers' Stockholm discussions were continued during the following months. At a Foreign Ministers' meeting in September it was agreed to set up a politico-military sub-committee, the so-called Defence Committee, with the task of investigating the prerequisites and possibilities of cooperation in defence between Denmark, Norway and Sweden. While the committee was sitting, the Danish and Norwegian Foreign Ministers participated in the UN General Asssembly in October. During their conversations with the US Secretary of State, George Marshall, they raised the important question of American supplies of arms to a Scandinavian alliance, and the economic conditions that would be attached to such deliveries. Marshall's answer was discouraging. It intimated that a neutral Scandinavan defence organisation should not rely on American military guarantees and, furthermore, that states which did not join the Atlantic alliance, then under discussion, would be of secondary importance in regard to the supply of arms. On the other hand, the American Government initiated, during the autumn, considerable diplomatic activity to convince the Scandinavian governments that it would be in their interest to join an Atlantic defence system.

In January 1949 the Defence Committee submitted its report. It highlighted the point of view that a common military effort would substantially increase the defensive power of the three countries through the widening of the strategic area, preparatory planning and the standardisation of equipment. The report emphasised however, that an absolute prerequisite was a substantial rearmament of Norway and Denmark. It was furthermore necessary to obtain important military equipment from countries outside the Scandinavian area and on favourable economic conditions.

Finally, the report discussed the general foreign, political and defence situation of the three countries. In spite of the strategic importance of the Scandinavian area, the authors did not take it for granted that it would automatically be involved in a conflict between the Great Powers. But they also emphasised that without military help from outside, a Scandinavian alliance would not be able to hold its ground for any length of time against an attacking Great Power.

This report formed the basis of discussions when the Prime Ministers, Foreign Ministers and Defence Ministers met at Karlstad in Sweden on 5–6 January 1949. The Swedes and the Danes each introduced drafts for a treaty, mutually binding on the participants to consider an attack on one of them as an attack on all. In case of a war which did not affect any of the countries, they should consult each other on how to prevent their territories from becoming involved.

Apart from a discussion of the divergent Norwegian and Swedish concepts of the relationship between a Scandinavian alliance and an Atlantic alliance, the main discussion at Karlstad was on how Denmark and Norway could be effectively re-armed. It was decided to investigate how arms could be obtained on favourable economic conditions without membership of the Atlantic bloc. The US Government, however, gave a negative answer to this question by a public announcement on 14 January, maintaining that military equipment would primarily be supplied to states which joined the United States in a collective defence organisation.

This actually decided the fate of the Scandinavian negotiations. During meetings at the end of January in Copenhagen and Oslo the divergent Norwegian and Swedish points of view were sharply spelt out. While Sweden, supported by Denmark, only wanted to ask Washington about the supply of armaments, the Norwegians also wanted to ask about the help to be expected from the United States in case of war. The Swedish aim was an independent Nordic alliance while Norway would only accept such an alliance combined with an association with the Atlantic organisation.

On this theme the negotiations ended, and the communiqué of 30 January 1949 from the Oslo meeting established the fact that it was impossible to reach agreement between all the Scandinavian countries on the conditions and consequences of an independent Nordic defence alliance. The day before, the Soviet Government had warned Norway against membership of a Western alliance. In the second week of February, Halvard Lange flew to Washington to discuss with the American Secretary of State, Dean Acheson, the conditions for Norwegian membership in the future North Atlantic Treaty Organisation.

Although under increasing pressure from the Liberal and Conservative Parties, the Danish Prime Minister, Hans Hedtoft, did not abandon his hope of a Nordic solution to Denmark's security problems. After the Oslo meeting, he approached the Swedish Prime Minister, Tage Erlander, suggesting a Swedish-Danish alliance, but all the leaders of the

Swedish political parties rejected this idea as unrealistic; as in 1863, the Danes' vision of a Scandinavian solution to their security problems faded away because of the hard facts of geography, national interests and external influence.

Denmark's search for a Nordic solution to its security problems finally came to nothing in January 1949. Simultaneously, negotiations were under way for an Atlantic defence alliance. The recent publication of American and British documents, and the opening of the Danish Foreign Office archives, have thrown new light on the impact of negotiations for an Atlantic defence alliance — which were going on simultaneously — on the Scandinavian discussions and the eventual Danish decision to choose the Atlantic solution for its national defence problems.

There were different motives for bringing the Scandinavian states into an all-embracing Western defence arrangement. Ideologically, their old democratic heritage made them natural allies of the Western nations. Of immediate interest to the Western powers, and in particular the United Kingdom and the United States, was their strategic importance. While the discussions preceding the Brussels Treaty of 17 March 1948 showed that the continental states were primarily interested in protecting Central Europe, the British Foreign Secretary, Ernest Bevin, took a wider view. Worried by Stalin's proposal to Finland at the end of February, and by rumours of a similar Soviet move towards the Oslo Government, he warned Norway not to set her foot on the slippery slope by sacrificing her right to conclude pacts with whomsoever she chose. Simultaneously, he sent the American Secretary of State, George Marshall, a message suggesting the probability of an imminent threat against Norway. Bevin maintained that he could not risk a Norwegian defection because it would involve the approaches to the Atlantic and would result in the collapse of the Scandinavian system. "This would in turn prejudice our chance of calling any halt to the relentless advance of Russia into Western Europe," he said. However, he rejected the idea of including Scandinavia in a Western European defence pact, and as far as is known he was the first responsible statesman to ventilate the idea of an Atlantic arrangement. In order to forestall a Soviet move and prevent Norway from "going under", he suggested "a regional Atlantic Approaches pact of mutual assistance in which all the countries directly threatened by a Russian move to the Atlantic could participate." Besides the United Kingdom and the United States, he mentioned as possible members of such a pact Iceland, Sweden, Norway and Denmark.

These feelers put out by Ernest Bevin were received with caution in Washington, but in June 1948 the Senate adopted the Vandenberg Resolution, authorising President Truman to participate in negotiations for a collective security pact with the Brussels Treaty powers (Belgium, France, Luxembourg, the Netherlands and the United Kingdom), as well as Canada and other states willing to accept the conditions in the UN Charter's Article 51 about individual and collective self-defence.

Norway's long Atlantic coast, which had been a hiding place for German warships during World War II, was obviously an important element in the decision to consider the possibilities of an Atlantic defence arrangement. But the strategic position of Denmark also came into the picture. The bases for the Soviet submarines were in the Baltic, and in order to get out into the Atlantic they had to pass through the Sound and the Belts and further on to Skagerrak, where the northern part of Jutland and southern Norway constituted the last barrier to the North Sea and the Atlantic. Vessels returning for repairs and provisions would, of course, encounter the same obstacles. The possession or control of Danish territory would consequently be important both to the Soviet Union and to the Anglo-Saxon naval powers.

But even more important to the United States was the Danish North Atlantic island, Greenland. When the Germans had occupied Denmark in 1940, the communications between Copenhagen and Greenland had been broken off, the local authorities establishing contacts with the Danish Minister in Washington, Henrik Kauffmann. As the Americans became more and more involved in the war in Europe, their interest in Greenland increased. In April 1941 Kauffmann, who had broken off his relations with the Government in Copenhagen and was acting as the agent of "Free Denmark" in the United States, concluded an agreement with the US Administration granting the United States rights to establish bases in Greenland.

Danish sovereignty over Greenland was explicitly recognised and its Article 10 stipulated that the Treaty should remain in force until the signatories agreed that the dangers to the peace and security of the American continent had ceased. There is no doubt that the Greenland bases and facilities were of great importance for the US war effort during the ensuing years.

After the liberation in 1945, the Danish Parliament ratified the Greenland Treaty, but later post-war Governments wished to abrogate it, and opened negotiations with Washington about US withdrawal from the island. The US Government's policy, however, was to seek to prolong the Treaty or to negotiate an alternative arrangement because of Greenland's importance after 1945 for the international meteorological service and the civilian air traffic between the United States and Europe. As the "cold war" developed, and a direct Soviet air attack on the United States became possible, the US Polar strategy included the important Greenland bases. When the Treaty of Rio de Janeiro was signed in 1949, Greenland was included with the American states in the area covered by the Treaty's mutual defence arrangements.

Under pressure from the political parties, the Danish Government continued to demand the abrogation of the 1941 Treaty, and one of the reasons for this was the desire to avoid Soviet suspicion or accusation of military co-operation with the only power feared by the Soviet Union. On the other hand, it became quite clear that Denmark, for economic reasons, was unable to take over the bases and meteorological installations on Greenland.

From the beginning of 1948, the more important question of a defence alliance overshadowed the Greenland negotiations, and at the end of March the Danish Government informed Washington that it did not intend to continue its demand for the abrogation of the 1941 Treaty. It was, in any event, becoming evident that the United States did not intend to leave Greenland; Denmark, on the other hand, wanted a *quid pro quo*.

The negotiations between the Brussels Treaty powers, the United States and Canada entered a decisive phase in June 1948, after the Senate had passed the Vandenberg Resolution. Among the important questions which were discussed was the widening of the group of states that would be offered membership in the Atlantic pact, and among those was Denmark. The views of the US authorities were divided. The Joint Chiefs of Staff considered Denmark's position as virtually hopeless and were critical of formal or informal guarantees to the Danes. On the other hand, as well as Greenland, the United States was interested in the Sound and the Belts. These divergent views were apparent in George Marshall's conversation with the Danish Foreign Minister, Gustav Rasmussen, on 5 October 1948. The American Secretary of State emphasised US interest in blocking the Belts "in the event of trouble", but also spoke "pointedly and frankly of the vulnerability of Denmark", which is described as the most exposed country in Western Europe. In a conversation with the Norwegian Foreign Minister, Lange, he characterised Denmark's position as "if anything, tragic."

The working group, which after July 1948 drafted what became the North Atlantic Treaty, agreed that Norway, Sweden and Denmark would be desirable and helpful participants. But according to Ambassador Achilles, a member of the group, "we worried lest Sweden might succeed in enticing the other Nordic countries into what we considered an inadequate Nordic defence agreement."

The positive attitude towards Danish participation in an Atlantic pact prevailed against the doubts, and the "stepping stones" concept, i.e. the necessity of including Greenland in the Atlantic defence system, may have been decisive. In January 1949 Henrik Kauffmann received the draft of the Atlantic Treaty. After the final collapse of Hedtoft's Nordic policy, he declared to the governing body of his Social Democratic Party that the Government could not take the responsibility of leading Denmark into isolated neutrality, and he received the Party's support for cooperation with "other democratic nations", i.e. participation in the Atlantic pact. This decision was also supported by the Liberal and Conservative Parties.

After the Foreign Minister's talk with Dean Acheson in Washington in the beginning of March, a majority in the Danish Parliament carried a resolution accepting Denmark's accession to the North Atlantic Treaty, and on 4 April Gustav Rasmussen signed the Treaty, together with eleven colleagues from the Western world. The most decisive step in Danish twentieth-century history had been taken. Every parliamentary election and opinion poll since 1949 has indicated that a majority of the Danish people continue to support membership of NATO.

THE UNCERTAIN MONTHS
British anxieties over the outcome of the negotiations

Alexander M. Rendel

Papers issued to the British Cabinet by Ernest Bevin, the Foreign Secretary, to guide their discussions on the Treaty were made public on 1 January 1980. Those memoranda published in 1949 confirm that the Treaty was only brought to birth after a period of the most anxious uncertainty, and that even within a few weeks of the signature it seemed possible that a wholly defective infant would finally appear.

Thus on 18 February 1949 Bevin issued a memorandum to the British Cabinet in advance of their meeting the following day, from which it is clear that the actual commitment which the NATO members were to make to one another — the wording of Article 5, the heart of the Treaty itself — could have been so seriously watered down as a result of a debate a week before in the United States Senate, that the British Government might have had to consider whether the resultant text would have been worse than no treaty at all.

Bevin begins his memorandum with the words: "My colleagues will have read in the press of the deplorable [*sic*] debate which took place in the American Senate on 14 February...." He then refers to three telegrams from Sir Oliver Franks, the British Ambassador in Washington, which evidently painted the Washington scene in the gloomiest colours, and comments that the debate, "I need hardly say, had a depressing effect on Mr Lange and has also no doubt encouraged the Soviet Government and their sympathisers all over the world."

Bevin had at the time been in contact and close personal alliance with Halvard Lange, the Norwegian Foreign Minister, and it is fair to say at once that the reluctance of some United States Senators to make any clear and strong commitment to their allies never shook Lange's own firm support for the alliance, and that, among Norway's political leaders, he remained throughout the one who more than any other was responsible for

Norway's eventual signature of the Treaty. But in spite of the considerable weight which he carried in the Norwegian Government, the position of Norway in mid-February 1949 was in real doubt, and there were still uncertainties about the Danish Government agreeing to sign a North Atlantic Treaty. Membership of both countries was regarded by their eventual allies as of great importance. Norwegian membership would mean that a Soviet advance into the Atlantic could at that time be easily contained with the available naval forces, while Danish membership was already of the greatest importance since it would bring Greenland within the area of the Treaty. Greenland, in some comments at the time, was said — perhaps loosely — to be essential for the defence of the United States and Canada.

Bevin's memorandum of 18 February 1949 makes it reasonably clear that he himself was at this point already determined that any treaty which provided some advance military planning between the Allies was better than none — however few teeth were apparently left in the wording of the Allies' mutual commitment after the mauling which it had just received in the United States Senate. His memorandum compares three possible drafts which the United States Government might back.

The first of these was the draft prepared by the Working Group of Ambassadors which had been meeting in Washington, representing the United States, Canada and the Brussels Treaty Powers (Britain, France and the Benelux countries), who had latterly also been joined by the Norwegian Ambassador. This committed all the Allies to use "military or other action" in the face of an attack upon any one of them.

The second draft, however, was a significantly modified version. It emerged during the United States Senate debate of 4 February in the light of Senators' preliminary objections to the first version. This second draft did not contain the words "military or other action", but retained the thought that an attack on one ally was an attack on all.

It was, however, in Bevin's view, much preferable to a third draft, which Senator Connally had publicly proposed "apparently", Bevin says, "at the suggestion of the State Department". This omitted not only the words "military or other action", but also the concept that all the Allies were committed by an attack upon any one of them, and furthermore provided that the action which an Ally might take was a matter for the individual judgment of that Ally itself.

Bevin was evidently appalled by this third version. He was no doubt worried, in particular, that if no firmer guarantee of support were forthcoming, the alliance might lose Norway and Denmark, but he was also evidently anxious to forestall and disarm objections which he expected from his Cabinet colleagues, some of whom, he thought, would arrive at the Cabinet meeting primed with lurid press accounts of the Senate debate. At any rate, his memorandum at this point seizes the bull by the horns — not so much to overthrow it, as to hold it in the position reached and prevent anything worse happening before Dean Acheson, who had recently been appointed the US Secretary of State, could come to the rescue.

Of this third weakest draft, Bevin therefore says: "Even in the event of this last version being accepted, however, it has been suggested by some Senators that it would not connote even a moral obligation on the part of the United States to go to war if a co-signatory of the Pact is the victim of recognised aggression." But he adds: "No doubt in such circumstances the United States *would* go to war, more especially of course if United States armed forces had previously engaged in hostilities. Mr Acheson is no doubt doing his best to wrestle with the Senators and we shall know the results of his efforts in a few days' time when the Ambassadors' committee reassembles."

This particular memorandum is certainly a good example of the roundabout way in which Bevin's thoughts so often moved nearer to the heart of the situation, preparing the ground before actually leading colleagues right to the central point which was, in fact, already clear enough to him. At any rate, after suggesting that the third draft, if accepted, would not in practice be as disastrous as its wording — or lack of stiff wording — might imply, he in effect advised the Cabinet not to spring forward with objections to the folly of some US Senators, while Acheson was trying to improve the situation. He was really saying that critical comments from the Allies at that stage would not help Acheson, but he was also preparing the ground for acceptance of the view that even if Acheson failed to get a better draft than the third version, then its weakness would not be sufficient in itself for any British rejection of the treaty. Although Bevin was evidently clear enough in his own mind that any treaty which allowed advance military planning was better that none, he was concentrating then upon tactics to obtain the best treaty possible.

Bevin's memorandum adds that the draft which had emerged from the Senate debate — the second of the three under discussion — was better than Senator Connally's proposal, since it retained the thought that an attack on any one ally was an attack on all, and because it did not specify that the determination of any action that any of the Allies might take was a matter for that Ally's individual judgment.

Bevin clearly had serious doubts whether a version satisfactory to the Europeans in including a clear and firm promise of military support would be acceptable to the Americans, and at this point he says: "If we can only get (c) [Senator Connally's proposal], we shall have to decide whether a pact so weakened is still worth signing."

At this point Bevin's thought turns aside, and he reminds the Cabinet of the further unsolved difficulties, such as the definition of the area of the Treaty; the question of whether Italy should be included, and whether Algeria should be within its scope. Of these problems, he says: "I need not here decide the further unsolved difficulties . . . since, though important, they are all fairly largely dependent on the solution adopted in regard to Article 5, and cannot be dealt with satisfactorily until that has been settled."

Bevin then returns to his real objective — to get Cabinet approval for even the greatly weakened commitment proposed by Senator Connally in the third draft, if no stiffer commitment was forthcoming from Washington, and also to decide how the British Government should play its hand meanwhile. Of Senator Connally's proposal, he says: "With the feeble version of Article 5 we should still presumably secure consultative machinery, and above all, the establishment of a military committee, which would be capable of drawing up plans and of dividing up the available arms among the signatory powers.

"By such means we could, of course, hope that, when and if aggression came, the operative clause [Article 5] would be of less importance, since under the various schemes and plans adopted by the military committee, a situation would have been created in which in fact the United States would not be able to avoid being involved in the conflict, whatever view the Senate took as to its technical right in regard to the declaration of war. In the end, I suppose, it is the existence of prepared common plans rather than of paper commitments, which usually proves effective in determining a Government to go to war in aid of an ally."

Bevin was therefore clearly determined that the British Cabinet should accept any of the three drafts under discussion, and was no doubt confident that he would succeed, but he was genuinely anxious that several of the European governments would, even at that late stage, back out. His memorandum goes on to state that he could not say what the views of other Brussels Treaty powers, "notably the French", would be. "The French", he declared, "have unhappy memories of being let down by the Americans in 1919, and it may be that the Parliamentary Opposition . . . would profit by the obvious lack of teeth in the Pact as drafted to cause the defeat of the Government if it accepted it and endeavoured to get it ratified by the National Assembly. On the other hand, a weakening of Article 5 would also mean a loosening of French and other European commitments in the event of a Soviet-American conflict in some other part of the world, and might therefore help certain elements in French opinion to accept the Pact."

Faced with these uncertainties, Bevin then turns to tactics, and argues that possibly it would be best for all the Brussels Treaty powers to tell the Americans that they could not guarantee the acceptance by their parliaments of "the Senate version" (the second draft). He suggests that they might aim "to get Acheson to get a version making it clear that military aid would be afforded by America to a co-signatory whom the United States Senate had itself recognised as a victim of aggression". This, Bevin argued, would give an assurance of military aid (real teeth) in the commitment, while making it plain that nothing could be done except with Senatorial consent.

In the light of these anxieties and uncertainties, British documents, although they are unfortunately seriously incomplete, are sufficient to confirm the debt which all the Allies owe to Dean Acheson, for within four days Bevin was, on 22 February 1949, sending a further memorandum to the Cabinet to inform them that Acheson had seen the Senate Foreign Relations Committee, and that there was "every reason to believe" that the

Committee would accept a version of Article 5 of the Treaty which would be acceptable to the Brussels Treaty powers. The latest draft, he emphasised, makes clear that the Treaty was a matter of mutual assistance and not merely of United States support for Europe. It contained as the operative words of the commitment "action including the use of armed force" in place of "military or other action".

This draft was declared in discussion by the British Cabinet to be an improvement even upon the original draft prepared by the Ambassadors' Working Group. The new formula did give each Party the right to determine what action was necessary "in accordance with established constitutional processes", and it was acknowledged by Bevin that this right might give rise to difficulties at some later date, "for it might", he wrote, "be invoked by unfriendly elements in the United States with a view to stressing the freedom of action of the parties in time of emergency." Nevertheless, it would be expedient, Bevin explained, to accept the Treaty as now drafted rather than run the risk of controversy in Congress.

Bevin, therefore, favoured an early signature of the Treaty on that basis, and on this and on a number of other points there was general agreement with him. When the text of the commitment was finally agreed on these lines, the British Government sent a warm message of appreciation to Bevin who was at the time in Washington.

Meanwhile, one potentially awkward question — whether Italy should be included in the Treaty or not — had solved itself. In the winter of 1948 and early months of 1949, Norway had been subjected to heavy pressure from the Soviet Union to hold herself neutral. In a memorandum to the Cabinet on 9 March, Bevin explained that this made it impossible, once Norway had decided to seek entry into the Alliance, to oppose her immediate acceptance, and the French then insisted with the utmost vigour that Italy must be treated no less favourably. Italy became, therefore, a member of the Alliance from the start, and the alternative of a Mediterranean Pact in which Italy might have played a leading role, and which was being vaguely mooted, was dropped for the time being.

This development, however, made it all the more important that Greece and Turkey should not be left out in the cold. Both wanted full membership, but were already protected by the United States through the Truman declaration of 12 March 1947, by the United Kingdom and, theoretically in the case of Turkey, by the Treaty of Mutual Assistance with France of 19 October 1939, though it was assumed from the French attitude that France no longer regarded the 1939 agreement as binding.

The Western Governments concerned therefore considered a draft declaration in which they would all affirm their interest in the integrity and independence of Greece and Turkey. Bevin would have accepted this and would also have included Iran if the United States had desired that, but at the time he had in mind separate defence agreements to cover the Mediterranean and the Middle East, and argued that a declaration to support Greece and Turkey would do nothing to buttress their security in practice and might seem to cast doubt on the effectiveness of the existing commitments.

It is interesting to consider the reasons why Greece and Turkey were not included in the Alliance at that stage. The record of the exchanges on this point is incomplete. One major reason in Bevin's mind was probably the reluctance of some of the Brussels Treaty powers and the Scandinavian Governments to accept a defence commitment in the Eastern Mediterranean. This could have delayed the signature of the Atlantic Treaty, and Bevin's main and most urgent objective at this time was to have it signed at the earliest possible date.

It should be added that even when a "most satisfactory" text was agreed and signed on 4 April 1949, there remained many unresolved problems which might still have gravely weakened the Alliance. During much of the Alliance's first year the planners remained anxiously doubtful whether any sensible regional defence system could be established. It seems almost incredible today how little organised the defence of Europe was in 1949, but in fact for much of that year the United States negotiators considered that they should not take part as full members in the regional bodies to be established under the Treaty except in the Atlantic and West Mediterranean, where only naval and air forces were involved. Simultaneously the French argued that the Treaty would be a dead letter unless the United States played a full part in the West European Regional Defence body which it had been agreed to establish. Nevertheless although the problems of military organisation and much else were still to face the Allies with many a hard-contested wrangle in the months and years ahead, there was never any doubt, after 4 April 1949, that that day ranked as an anchor date in the history of Western democracy.

A CHANGE OF HEART

Concerns behind the discussion in France

Claude Delmas

The different members of the North Atlantic Alliance all contributed in their own way and from the very start to the negotiations which led to the construction of the common edifice. France's contribution is of particular interest firstly, because of the change of heart represented by the adoption of the Atlantic option in place of the anti-German tendencies of previous years and, secondly, because of the contrast between its stand in 1948-9 and its withdrawal from the military structure of NATO in 1967.

France's participation in the creation of the Alliance was constantly hampered by internal political difficulties. On 4 May 1947 the Prime Minister, the Socialist Paul Ramadier, had to dismiss from the Government Maurice Thorez and three other Communists (one of whom was the Minister of Defence, a fact which had compelled the General Staff to take secret precautionary measures against sabotage and subversion). On 22 September 1947, at the Szklarska-Boreba Conference (at which the Cominform was to be set up), Andrei Zhdanov ordered Maurice Thorez (and Palmiro Togliatti) to organise opposition to the "infamous Marshall Plan".

Disorder immediately increased and strikes took on insurrectionary proportions. A Socialist, Jules Moch, was given the task of safeguarding freedom and order. In the meantime, the Gaullist movement had been formed. The Rassemblement du Peuple Français, created on 14 April, decided on a policy of systematic opposition. The Government, which was a coalition of centre parties (Socialists, Christian Democrats, Radicals) was constantly under threat from an unnatural alliance of Communists and Gaullists at a time when it was grappling with particularly serious problems: Communist agitation and subversion, a financial crisis, the question of Government funds for church schools, the consequences of the Malagasy uprising, the beginning of the war in Indo-China and so on. It was against this background that it took part in the negotiations which were to lead to the Washington Treaty. Its initiative was warmly supported by public opinion which, while it had remained relatively indifferent to the creation of the Cominform, was outraged by the events of February 1948 in Prague.

On 4 March 1948, with the Brussels Treaty on the horizon, Georges Bidault, the Foreign Minister, sent a message to General Marshall in which he said, "The time has come to tighten collaboration in the political field, and as soon as possible in the military field, between the New World and the Old, joined as they are by their commitment to the only worthwhile civilisation." Even before the signature of the Brussels Treaty, therefore, Bidault was thinking in Transatlantic terms. On the day of its signature on 17 March, he said, "Peace has always been undermined by isolation, fear and mistrust; together we shall overcome these three dangers."

On 3 April, in submitting the Treaty to the National Assembly, he added: "I would remind those who have spoken of a Western bloc directed against other countries that there are already fifteen treaties between the countries of Eastern and Central Europe. Western Europe is surely entitled, in the pursuit of liberty, to act as others have already done elsewhere, not against others but like others", which was tantamount to holding up the Brussels Treaty as an act of self-defence.

There was an obstacle, however, to the complete alignment of the policies of France, the United States and the United Kingdom. The Communist coup in Prague had illustrated the true gravity of the Soviet threat. France had undoubtedly realised that, given Moscow's attitude, there was no possibility of a solution to the German problem and it was, moreover, obvious that with every quadripartite meeting new problems sprang up faster than the old ones could be solved. However, on the two issues of reparations and international control of the Ruhr, it stood by the position which it had adopted in 1945 and which contained an undeniable contradiction, namely a streak of anti-Germanism combined with the desire to see Germany reinstated in the concert of nations.

As from July 1948, France's attitude became more flexible. On 28 September Robert Schuman came out in favour of the reconstitution of Germany on a federal basis, and on 18 October he agreed to the principle of a merger between the three Western zones.

This difference of opinion regarding the solution to the German problem had not prevented the French Government from playing an active part in setting up the bodies provided for in the Brussels Treaty, a Treaty which had been greeted by President Harry Truman, when addressing Congress on 17 March, with the words, "I am sure that the determination of the free countries of Europe to protect themselves will be matched by an equal determination on our part to help them to do so."

France played a particularly active part in all the meetings which were, on the one hand, to make "Western Union" a reality and, on the other, to give an Atlantic dimension to the system of collective security set up under the Brussels Treaty.

By the spring of 1948, it was felt in Paris that while the "Western Union" could play an important political role, it did not go far enough in the security field. The Soviet Union's reaction had moreover been revealing; realising that the forces of the signatory states, even when united, amounted to little when compared with the Red Army (which was still on a wartime footing), it immediately set about organising the Berlin blockade.

Western Europe was determined to defend itself, but could not do so alone. Consequently in an official statement issued in Paris on 26 October 1948, the five Foreign Ministers of the Brussels Treaty states announced a complete identity of views as regards the principle of an Atlantic defence treaty and the steps to be taken in this connection. At Melle, in Western France, on 8 November, the military leaders of the "Western Union" and the Commander of the United States Forces in Germany met to consider the concrete provisions of such an "Atlantic Treaty". The French Government then drew up an initial draft providing for a 50-year (as in the case of the Brussels Treaty) military alliance which it submitted to the Standing Committee of the Brussels Treaty countries, meeting in London, on 11 November. A final draft was put together on the 26th, approved by the Governments and sent to Washington on 29 November.

A first meeting between the Ambassadors and the State Department to examine this draft took place on 4 December.

France was represented by its Ambassador, Henri Bonnet, assisted by its Minister Counsellor, Armand Bérard, and Arnaud Wapler, a Counsellor at the Embassy. On the same day Robert Schuman officially informed the Foreign Affairs Commission of the National Assembly of these negotiations.

The negotiations were still in progress when, on 25 February 1949, the Prime Minister, Henri Queuille, made an important statement to the United States Press Agency: "The United States must never allow France and Western Europe to be invaded by Russia as they have been by Germany," he said. But "France, on the outposts of Europe, cannot hold out alone. If sufficient forces could be relied on to prevent the Russian army from crossing the Elbe, the civilisation of Europe would be safe. A fortnight after an invasion, it would be lost." The meaning to be read into this statement was that France was not only concerned about the outcome of the battle, but also about the position of the front line in the event of such a battle. In the French view, this must be east of the Rhine, implying substantial United States military aid but also what was subsequently to be known as the forward strategy, as well as the inclusion of the three Western zones of Germany in Western strategic planning.

With this statement by Queuille, France gave its implicit support to a merger between the three zones which was to lead to the "Bonn Basic Law", i.e. the creation of the Federal Republic. From that moment, Germany was no longer a "problem" for which a four-power solution was needed, but had returned to the international scene as a partner.

The final version of the Treaty reached Paris on 8 March. On 9 March it was read out by Robert Schuman to the Cabinet, which on 16 March approved the accession of Denmark, Iceland, Italy, Norway and Portugal. The text was published on 18 March, and referred to by Schuman in a broadcast statement in the following terms: "We have today obtained what we hoped for in vain between the wars: the United States have recognised that there can be no peace or security for America if Europe is in danger. . . . The countries of Eastern Europe are united through a network of 23 bilateral treaties of assistance. These forces are subordinated to an ideology which does not attempt to hide its expansionist aims and which, since 1947, has been able to rely on that powerful instrument, the Cominform."

A change of heart: the discussions in France

As was to be expected, the Communists pulled out all the stops. On 22 February, Maurice Thorez was already railing against "an anti-Soviet war" and proclaiming that "the people of France will support the Red Army." The Communist Party launched an intensive campaign on the theme "France will not fight the Soviet Union" and mobilised all its organisations "for peace". Vincent Auriol, President of the Republic, decided that it was time to intervene: "I cannot allow it to be put about that there are people in France preparing to drag our country into a war of aggression" and in an allusion to the Eastern bloc: "We cannot allow ourselves to remain isolated when others are uniting."

On 4 April in Washington, after signing the North Atlantic Treaty on behalf of France, Robert Schuman stated: "We must join together to stamp out any idea of aggression by making it increasingly dangerous for the aggressor. Only a potential aggressor could have any grounds for considering itself threatened by the Treaty." These words heralded the policy of deterrence which was to follow.

The last step was to get the Treaty through the National Assembly. On 17 May, the Government tabled a bill authorising the President of the Republic to ratify the Treaty. On 2 July, the Foreign Affairs Commission rejected by 20 votes to 13 a motion by François Billoux (Communist) calling on the Assembly to refuse to discuss the Treaty on the grounds that it was an instrument of war against the USSR and evidence of "plans to increase colonial oppression and place France under the orders of United States imperialism". The Commission then approved the bill after amending its sole article. This article was then drafted as follows: " The President of the Republic is hereby authorised to ratify the North Atlantic Treaty concluded in Washington on 4 April 1949. The agreement provided for in Article 10, whereby States not party to the Treaty may be invited to accede thereto, shall not be given by the President of the Republic unless so authorised by a law to this effect."

The significance of this amendment was quite clear: accession of a new member could not be agreed to by the Government alone but would have to be approved by the National Assembly, thus ruling out the possibility of admitting Germany without parliamentary control.

On 11 July René Mayer, on behalf of the Foreign Affairs Commission, tabled a report which has become a landmark in the annals of French parliamentary history.* He outlined its main arguments and conclusions to the National Assembly on 22 July, during which he said, "We did not fight a war against one dictatorship to permit another to take over Europe." And, in a reference to the Treaty itself, "We can only feel pity for those who see the Treaty as the fruit of reactionary plotting and who, in turning down Marshall aid, the Atlantic Pact and European Union, are no doubt hoping to perpetuate a status quo which leaves Europe open to expansionist and totalitarian ventures."

There were cries of treason from the Communists. Their spokesman François Billoux described the Treaty as the "most monstrous act in the history of reactionary governments," "a Holy Alliance against the freedom of individuals and of nations" and "the new version of the anti-Comintern treaty." It would have needed more objective arguments to impress Robert Schuman, who found it perfectly easy to remind his listeners that the Washington Treaty was very much more recent than the Soviet system of alliances and the monolithic structure erected by the Communist bloc forces. "Could we remain supine in the face of this Cold War, waged by a central authority, backed by military forces far more powerful than our own countries? We would have failed in our duty," he said. "It would be unwise of us to overlook the preamble and the first two articles. They go well beyond the usual well-turned phrases. Indeed on examination they will be seen to express a desire for cooperation and understanding which far exceeds concern for mere military defence . . . I am aware of the dogged opposition of the Communist Party. This is hardly surprising since, according to the Party, France can only opt for one or two things, isolation or allegiance to the Soviet Union. We want neither."

The vote came at the end of a debate which the Communists deliberately stirred up to fever pitch (some members even came to blows). Finally, on 27 July at 8.00 a.m., the bill was adopted by 395 votes to 189. Those who voted against were the 168 Communists, the eight "Progressives" (pro-Communists, not

---

*It was eventually published by Presses Universitaires de France.

members of the Party), the six Deputies from the Rassemblement Démocratique Africain, the four Algerian Deputies who supported Messali Hadj, two Christian Democrats and one "Independent". There were 15 deliberate abstentions including the 13 "Overseas Independents". The Conseil de la République adopted the bill by a very large majority: 284 votes to 20. Ratification of the North Atlantic Treaty could thus go ahead.

The opposition then took up the issue of German rearmament, but times had changed. The anti-German aspect of French foreign policy was becoming a thing of the past and a few months later Robert Schuman, building on the basis of Franco-German reconciliation, was to lay the foundations of the European edifice. Neutralism was confined to a few groups of intellectuals, and Atlanticism became the cornerstone of French diplomacy. Three men had contributed to this turn of events: Georges Bidault, the former Chairman of the National Resistance Council; René Mayer, whose only son had been killed in Alsace early in 1945, and Robert Schuman, born in Luxembourg of Lorraine parents. Each of these men, in his own way, was a symbol, standing for France's determination to reject totalitarianism and, in consequence, to join with those other nations which, having been formed by the same civilisation, were equally keen to ensure the survival of freedom. In this respect France played an active part in the genesis of what was to be an historic act, namely the creation of the Atlantic Alliance.

THE PULL OF THE CONTINENT
Portugal opts for a European as well as an Atlantic role

Albano Nogueira

When, early in 1949, Portugal was approached to participate in the North Atlantic Treaty, then in the making, Prime Minister Salazar had every reason to receive the proposal with much personal gratification. The exclusion of Portugal from the Treaty would surely have been, from the point of view of the best interests of the Allies, highly undesirable, not to say inconceivable. But the Second World War had ended only four years before, which was perhaps not long enough for any possible sins, detrimental to the Allied cause, which some politicians in the Western democracies might still ascribe to the Portuguese Government, to be forgotten and forgiven. That this was a defective — because a too radical — judgement to pass on the Government is not something we can discuss here.

The state of affairs in Europe in 1949, when the Western Allied leaders came to the conclusion that the wide gulf in their relations with the USSR could not be bridged in the foreseeable future, gave Salazar the melancholy right to proclaim how accurately he had forecast the development of European policies. In the same way, he could also justify the correctness of the "situation" of neutrality which he had declared at the start of the war — a position he had so skilfully reconciled with Portuguese commitments under the terms of the alliance with Britain when the latter had asked for facilities. These had been promptly granted in the Azores — incidentally the first test of their importance in any war involving North America and Europe.

As far back as May 1944, a year almost to the day before the end of the war in Europe, Salazar said to the Congress of the National Union (the only political organisation then allowed in the country) that, to the extent that the USA would "consider it to be in their interest or their duty to help Europe to rise again", the Atlantic area was liable to be "one of the major centres of world politics". He reiterated this cogent statement on 8 May

1945 — the day after Germany's acceptance of the surrender terms — when he said to the National Assembly that as the result of the victory of Britain and the USA, the centre of gravity of international politics would be transferred to the Atlantic. In an aside, he significantly hinted at a Portuguese interest in this development.

Some ten days later, he found it appropriate to go further, saying that the Second World War had been "the last time in which we [the Portuguese] could, and ought to, be neutrals in a European conflagration". These were grave words from a man whose political philosophy and very personal reading of history led him, time and again, to stress the idea of his country being historically alien to intra-European conflicts. These were not the illuminating utterances of a prophet, but rather the lucid judgement of a dispassionate observer of events and political trends, which pointed to a USSR growing away from the West and consolidating its hegemony over its neighbours. The word *hegemony*, now so fashionable, was used by him then in that context.

It cannot pass unnoticed that, in his remarks of 1944 and 1945, Salazar had raised, albeit in an embryonic form, the idea that the Western Allies would soon be compelled to work out some sort of deterrent to Soviet expansionist aggression, in the shape of a collective effort or joint venture. By showing his sympathy with the proposed North Atlantic Treaty four years later, Salazar looked at it, not primarily in terms of power politics, but as a moral duty to cooperate in the defence of a threatened civilisation. No wonder, then, that in the Report of the Corporative Chamber to the National Assembly, recommending ratification of the Treaty, the evocative word *crusade* was used. Thus special emphasis was given to the high aims to be pursued, as opposed to the confusing specificity of the multiple political and doctrinal assumptions in the text.

Coming down to earth, the Portuguese authorities were not entirely satisfied with the text as a whole. As was to be expected, the Prime Minister was not prepared to accept it without reservations. In his view, some of the objectionable sections could easily have been avoided, and he noted acidly "the *hesitation* of the doctrine, the *fluidity* of the precepts, the *imprecision* of some of the formulae" (author's emphasis); above all, the Preamble was "manifestly unfortunate", since "the emptiness or the imprecision of worn-out and disturbing formulae" might lead to differences of interpretation. It would be preferable if assertions "more in conformity with the agreed principles of a civilisation that ought to be defended" were to be substituted for the merely negative anti-Communist content of those formulae. That approach, he admitted with mild sarcasm, would indeed be improbable because "some serious consequences of individualistic liberalism" had to be handled with indulgence. Moreover, vain "attempts are made to reconcile freedom and socialism", and "spiritual energies are wasted in attempting an identity of contrarieties through the conciliation of Communism and Christianity." Nevertheless, responding with forgiving equanimity to those who had given him a clean bill of political acceptability, a conciliatory, indulgent leader expressed his ultimate consent: "Be that as it may, we feel bound by the obligations of the Treaty and by its general aims — not (in any way) by a doctrinal affirmation pointing to the uniformity of political regimes, of whose virtues in our country we have learnt enough."

All things considered, both those who invited Portugal and the Portuguese Government itself undoubtedly performed, in this impossible consensus of opinions, an act of high statesmanship in tacitly ignoring points of friction and resorting simply to a *selective reading* of the principles and assumptions set forth in the "manifestly unfortunate preamble".

It is not surprising that Salazar took a strongly negative view of Spain being excluded from the Treaty. Portugal, he emphasised, "was not conditioned by any engagement based on the political systems prevailing in the two countries, nor on any sort of political solidarity, which in fact does not exist." Portugal was only concerned with the "conciliation of peninsular interests and their integration in the area of European interests". And he concluded by stating that the Peninsula ought to be considered "as a whole" in the context of Western defence; the "value and significance of the Portuguese accession [to the Treaty] are different should Spain be included in it or not; if not, they will vary according to the policy she follows in a conflict requiring Portuguese involvement under the provisions of the Treaty."

Sensitive to the point of obduracy over accepting any

commitments which, because of their ambiguity or fluidity, might jeopardise the inalienable and supreme right of national sovereignty, the Portuguese Government made known, in an official statement released to the Press by the Portuguese Embassy in Washington, its firm decision "not to accede to any international engagement which might tie [it] to the obligation of granting, in peacetime, military bases to a foreign power" "Even less", the document continued, "should the decision on such a course depend upon the mechanics of an agreement in which the views of other powers might prevail." Such clear words hardly need further elaboration. However, it should be added that the Portuguese position did not denote merely the legalistic acceptance of a juridical concept, but was above all a precautionary step by a statesman who had lived out the war years under the actual threat of possible foreign action transgressing his country's sovereignty.

It was common sense that Portugal's contribution to the Alliance would derive mainly from its geographical location, in Europe to some extent, but first and foremost in the Azores. A small country, of little importance in terms of military power, and not particularly important as a bridgehead in Europe, it was of prime importance as a provider of infrastructure in the Azores. This was why the Portuguese Government understood it to be its duty to safeguard sovereignty and why, consequently, it asked for clarification on this point — although it stressed that its reassuring assumption on this point was substantiated in the Treaty itself. Afterwards, Salazar remarked obliquely in the National Assembly that "utilisation in peacetime of strategic points [*points d'appui*] in Portuguese territory" would *per se* be out of the question, since "we had firmly made up our minds not to allow undesirable servitudes" (which would arise from such utilisation). But it was well understood that United States support had become "necessary to the security of the countries bordering the North Atlantic", and conversely that "Atlantic positions had become necessary to American defence." And that, of course, was the important point to assume.

For many in Portugal (including Salazar) historical evidence indicated that Portuguese involvements in intra-European conflicts had always run contrary to the country's interests. Some interpreted this as confirmation of a geopolitical thesis, whereby

Portugal is an Atlantic and not a European country. This Portuguese "Atlanticist vocation", together with the lessons of history already referred to, made it clear that Portugal would be extremely reluctant to accept unreservedly any commitment to join in an intra-European conflict under the obligations stipulated in the Treaty. It is of some interest that the Government took a pragmatic view of the matter, relying more on facts than on taking exception to undefined principles. "It seems improbable", said Salazar, "that in the next twenty years [after which period any member-country could withdraw from the Alliance] a conflict might occur among the Allies." But, he added, "we could not commit ourselves to intervene in intimate European skirmishes arising out of the peace conditions, in the formulation of which we took no part." (As is well known, Salazar believed that a negotiated peace would have led, in terms of continental strength, to a more effective distribution of power in Europe than that eventually obtained by unconditional surrender.) Now, only "an eruption from the East" remained a possibility, and should "such a great cataclysm" be "imminent", it would be "our duty to come forward with our contribution to avoid it".

According to Article 8 of the Treaty, the compatibility of the commitments now assumed by its members with any prior ones still in force had to be openly declared by them. For Portugal the point involved something deeper than a simple declaration, because the Treaty, considered as a coherent whole, might bring about a significant dislocation in the traditional vectors of her foreign policy. What was at stake clearly went far beyond the scope of obligations previously assumed *vis-à-vis* other countries. Hence the question which Salazar put to himself in public: "Could the Portuguese adherence to the Treaty be, in some measure, regarded as a deviation from our traditional external policy?" He answered it more by implication than in clear words. Characteristically avoiding a choice between a sharp "no" and a straight "yes", he placated the doubtful and the hesitant by actually accepting the Treaty.

The Portuguese international commitments, in the political area, that had to be taken into account were only two: the centuries-old alliance with Britain dating from 1373, and the 1939 Treaty of friendship and non-aggression with Spain, and its Additional Protocol of 1940. Neither seemed to interfere with the Portuguese acceptance of the new Treaty.

The alliance with Britain had been for Portugal the unchanged basis of a policy which, in Salazar's words, was guided by the necessity to make the maritime routes secure, which were "of fundamental importance" to Britain and an "essential element of Portuguese life, in continental Portugal as well as in the other parts of the nation, which are spread throughout the world". It had lasted so long, even though strained from time to time by very serious crises, because of the flexibility with which it had been interpreted, according to the requirements of changing historical events. Now the USA had come to the forefront in assuming new commitments both in the Atlantic area and in Europe, but this would in no way clash with, or weaken, the provisions of the alliance with Britain. On the contrary, by making a decisive contribution to assuring the security of the Atlantic, which Britain had upheld hitherto, the old links could only be strengthened by such a *complementary* reinforcement. Certainly, regarding Europe or rather the European territory, no problem would arise because the Portuguese position was quite clear: an attack by the East, as Salazar summed it up, "should be taken as against all [the members of the North Atlantic Treaty], and against the principles of culture and civilisation they represent."

The Treaty with Spain of 1939 and its Additional Protocol of 1940, known then as the Iberian Pact, concluded under the threat — soon to be the reality — of war in Europe, was of a different nature and scope. While it was no doubt motivated by solidarity as between two regimes of similar political hue, its terms have nevertheless been grossly distorted by inadequate, if not deliberately biased, interpretations. The Treaty was clearly a regular pact of friendship and non-agression, so framed that it far exceeded the transient state of affairs then prevailing in the two countries. When the project for a North Atlantic Treaty was made public in 1949, Spain declared herself ready and willing to be a party to it, should an invitation be forthcoming. This clearly implied that accession to the Treaty was compatible with her obligations under the Iberian Pact. Portugal decided, notwithstanding, to open talks with Spain on the matter of Portuguese accession, and the fact that Spain raised no objection had to be interpreted merely as acknowledging Portugal's gesture of good-neighbourliness, and not, of course, as the expression of an unnecessary consent.

To sum up, it seems appropriate to recall the question posed in 1949 by Salazar as to the threat to traditional Portuguese foreign policy which the North Atlantic Treaty might embody. Thirty-five years later, it can no longer be doubted that the Atlantic has gained a wider dimension in international politics, while the political division which Europe was undergoing gave those of its countries that claimed to be the depositories of genuine European values a clear sense that the time when they could live apart as a mere conglomerate was past. It was in this context that the North Atlantic Treaty was born, assuring American protection for European weakness, and it was as a means of defence against, and protection from, Communist hegemony that Portugal accepted it. In some subtle way Portugal saw the Treaty as a powerful weapon for defending moral values and at the same time for preserving the independence of the Western countries.

The Treaty had worked out an improbable synthesis between Portugal's Atlantic vocation — or, put another way, her obvious historical, manifest destiny — and a necessary, even if reluctantly admitted, approximation to remaining European and Christian values, and to that extent Portugal had every reason to be satisfied with the results. It is not certain whether Salazar's prompt acceptance of the Treaty meant that he then foresaw the entry of Europe into a new era — a new historical context — of wider breadth, and in the process of beginning a search for an inner unity from which it could derive strength and an identity.

If there are any still alive among those who in 1949 opened to Portugal the paths of the future, they would realise how transient the constants of history can in fact be. Salazar lived long enough to contemplate, from his country on the periphery of Europe, the growth of a Europe attempting to find a means of survival. In that task the North Atlantic Pact played an essential part for Portugal, since it was the first step which the Portuguese had taken on the road leading to a continent on which, because of their fascination with the sea, they had previously turned their backs. A partner in NATO, the OECD, EFTA, and the Council of Europe, Portugal is now negotiating full membership with the European Economic Community. It is not difficult to figure out that none of these subsequent initiatives would have had Salazar's blessing, thus illustrating how important are the vagaries of historical events when one is re-assessing what was previously taken as fate, vocation or mission. Salazar's well-known and (to his credit) always undisguised scepticism as to the viability of an organised Europe was not due to an unfulfilled desire to achieve it. It was due rather to the apprehension that attempts to achieve it could eventually succeed.

THE ART OF THE ALMOST IMPOSSIBLE
Unwavering Canadian support for the emerging Atlantic Alliance

Escott Reid

We who participated in the twelve-month-long negotiation of the North Atlantic Treaty, which began on 22 March 1948, know that there might have been a better treaty, there might have been a worse treaty, there might have been no treaty. Indeed it seems to me, looking back at the negotiations from a distance of more than thirty years, that an intelligent observer at the beginning of 1948 would have said that a military alliance between the United States, Britain and other Western European countries was impossible: the inertia, scepticism and opposition in the United States, France and Belgium were too great to be overcome.

The Council of Foreign Ministers which had begun its sessions in London on 25 November 1947 broke up in complete disagreement on 15 December. That evening Ernest Bevin, the British Foreign Secretary, said to General Marshall, the United States Secretary of State: "There is no chance that the Soviet Union will deal with the West on any reasonable terms in the foreseeable future. The salvation of the West depends on the formation of some form of union, formal or informal in character, in Western Europe backed by the United States and the Dominions — such a mobilisation of moral and material force as will inspire confidence and energy within, and respect elsewhere." At Marshall's request, John Hickerson of the State Department went to the British Foreign Office the next day to try to secure an elucidation of Bevin's ideas. The official he talked to (presumably Gladwyn Jebb) said that Bevin envisioned "two security arrangements, one a small tight circle including a treaty engagement between the US, the Benelux countries and France; surrounding that, a larger circle with somewhat lesser commitments but still commitments in treaty form bringing in the US and Canada also."

Marshall immediately said to Hickerson that United States participation in a military guarantee was not possible. United

States participation would have to be confined to supplying material assistance to the members of a Western European security pact. Moreover his "Marshall Plan" speech was only six months old, the Europeans were still arguing over who was to get United States aid and how much, and Congress had still to authorise, let alone appropriate, any aid at all. When Marshall returned to the United States, he found that his Under Secretary, Robert Lovett, agreed with him in rejecting, as at least premature, any idea of a military alliance between the United States, Britain and other Western European countries. Marshall and Lovett were supported in their opposition to United States participation in a military alliance by two of the most able and influential senior officers of the State Department, George Kennan and Charles Bohlen. They were strategically well placed in the Department to bring their influence to bear since they were the acknowledged principal experts on the Soviet Union, and Bohlen was also in charge of relations between the Department and Congress. Kennan was Director of the Policy Planning Staff; Bohlen was Counsellor of the Department. As such, they were more senior in the hierarchy than the two principal advocates in the State Department of a military alliance, Hickerson and Theodore Achilles. Hickerson was Director of the Office of European Affairs, Achilles the Director of the Division of Western European Affairs.

At the beginning of May 1948, a month after the successful termination of the top-secret discussions on the Treaty between the United States, Britain and Canada, we in Ottawa learned that Bohlen was contending that an Atlantic pact would cause undue provocation to the Soviet Union and that Congress would be unwilling to undertake so far-reaching a commitment in an election year. Kennan was reported to us as having said that he had discussed the matter at length with Bohlen and that they saw eye to eye. It was unthinkable that the United States would stand idly by if the Soviet Union were to make an aggressive move against any country of Europe; the Russians were fully aware of this, and the best deterrent to action by them was not an Atlantic pact but the provision of military supplies to Western Europe by the United States, accompanied by a resolution of the Senate. The British Embassy in Washington received at this time a message from Bevin to Marshall putting the arguments for a North Atlantic treaty, but it was so shaken by the opposition of Bohlen and Kennan that it was inclined not to present the message since it might merely produce a reply indicating the United States' unwillingness to conclude such a treaty.

Towards the end of June, eleven days before the opening on 6 July of the six-power talks on the Treaty (the United States, Britain, France, Belgium, the Netherlands and Canada) Hume Wrong, the well-informed Canadian Ambassador in Washington, reported to the Canadian Government, "It is still uncertain whether the State Department is fully convinced that a treaty commitment is desirable. . . . I think that Lovett is inclined to be sympathetic but not fully satisfied on the treaty proposal. Bohlen still tends to oppose it, while Kennan appears to be converted. Hickerson is the staunchest advocate." On 9 July, the State Department committed itself to a treaty but it was not until more than seven months later that the two leading members of the Senate Foreign Relations Committee could be persuaded to support an effective treaty.

On 14 February 1949, a month before the final text of the Treaty was made public, Tom Connally, the Democratic Chairman of the Senate Foreign Relations Committee, declared in a speech in the Senate that he did "not believe in giving carte blanche assurance to [the nations of Europe] — do everything you want to do, you need not worry, as soon as anything happens we will come over and fight your quarrel for you". We should not "blindfold ourselves and make a commitment now to enter every war that may occur in the next ten years, and send our boys and resources to Europe to fight. We cannot . . . be Sir Galahads, and every time we hear a gun fired, plunge into war and take sides without knowing what we are doing, and without knowing the issues involved. There are many people, and we have found them in government and elsewhere, who would favour automatically going to war, which would mean letting European nations declare war and letting us fight."

Immediately after this speech, Connally and Arthur Vandenberg, who had just ceased to be the Republican Chairman of the Foreign Relations Committee, told Dean Acheson, the Secretary of State, that Article 5, the pledge article, should not refer to the possibility of the signatories of the Treaty giving military assistance, and that the Article should make plain that each of the Allies would determine individually the type of action it would take to assist an Ally that had been attacked.

Robert Schuman, the French Foreign Minister, told the Canadian Ambassador to France that the language proposed by Connally for the pledge was meaningless and unacceptable and that he would not sign any treaty which was not *"sérieux et complet"*.

The Canadian Government, on 17 February, authorised its Ambassador in Washington to inform the State Department that

> . . . the purposes of the Treaty are not going to be fulfilled by an undertaking which is so watered down that it does not create even a moral obligation to take effective action, but is put forward as a charitable donation from the United States. This is reducing the proposed North Atlantic Treaty almost to the level of a Kellogg-Briand peace pact. If there it no satisfactory pledge in the Treaty, and if the Treaty is interpreted by the Senate merely as a mechanism for getting the European states out of difficulties which really don't concern the United States directly, then its value is greatly reduced and we might have to re-examine our whole position. It might be that, in the light of such re-examination, we will be compelled to decide that the Canadian national interest involved in this kind of treaty, interpreted in this way by United States opinion, is not sufficiently direct and immediate to warrant the government recommending to Parliament our adherence to it. We would do this, of course, with the greatest regret, but we might, in the circumstances, conclude that it is better to have no treaty at all than to have a treaty which is so weak and ambiguous as to be meaningless and therefore mischievous, especially since the conclusion of such a treaty might render less likely the conclusion of a really effective arrangement in the future.

These are indications of the depth and extent of the opposition in the United States to a North Atlantic military alliance. What about France in 1948?

The French Government, as late as the middle of August 1949, kept insisting at the six-power talks on the Treaty that what was important was not an Atlantic treaty but two concrete and immediate requirements: the provision of military equipment from the United States to France and commitment by the United States to co-operate to stem a Soviet attack on Western

Europe as far east in Germany as possible. Lovett told Lester Pearson, then the Under-Secretary of State for External Affairs in Ottawa, that Henri Bonnet, the French Ambassador, had called on Marshall in mid-August and had suggested that the French would only accept an Atlantic security pact on the following three conditions: "Unity of command at once; immediate movement of US military supplies to France; immediate movement of US military personnel to France." According to Lovett, this had had "such an irritating effect on General Marshall that . . . he felt like calling off the Atlantic pact negotiations at once." On 24 August, Lovett sent a telegram to the US Ambassador in Paris beginning, "The French are in our hair."

The French views were apparently shared in July and August 1948 by Paul-Henri Spaak, Belgium's Prime Minister and Foreign Minister. The Netherlands Foreign Minister told the Canadian Ambassador at the Hague that, at the meeting of Foreign Ministers of the Brussels powers which had taken place there on 19 and 20 July 1948, the Belgian as well as the French delegates had said, concerning the talks in Washington, "Their governments were, for the moment, less interested in a long-term policy than in coping with any emergency in the near future."

Two weeks later, we in Ottawa learned that Spaak had said that he thought the North Atlantic security pact was premature and that the provocative effect on the Russians was not justifiable if one took into account the military potential of the North Atlantic countries. It was indeed possible that Spaak, a future Secretary-General of NATO, was at this time turning over in his mind whether an armed neutrality of West European countries might not be preferable to a North Atlantic treaty.

The Belgian Ambassador in Ottawa sounded out Lester Pearson on this in the middle of July. The Ambassador said that he was "speaking entirely personally and not necessarily reflecting in any way the views of his government", a formula customarily employed by an ambassador when he has received an instruction to sound out the government to which he is accredited without committing his own government. He expounded his "personal" ideas as follows: in two or three years, a rearmed and united group of West European countries might well be able to maintain their neutrality against the Soviet Union in the event of war between the Soviet Union and the United States. In any such West European combination, it would be essential to include West Germany (and, if possible, the whole of Germany), for the Germans could provide the strongest military forces to be used against the Russians. Such an armed neutrality would serve the interests of the United States since, if a conflict between the Soviet Union and the United States automatically involved West European countries, the United States might have to come to the help of these countries on land and thereby weaken its air attack on Russia which was the only way at this time by which the Russians could be defeated.

What was it that made possible the birth of the North Atlantic alliance in spite of the inertia and scepticism and opposition in Washington, Paris and Brussels? I shall not try to make a complete list of the factors that came into play. I shall discuss only four of them: India, Norway, the State Department and Canada.

The Canadian Ambassador in Washington reported to Ottawa at the end of May 1948 that the two main difficulties in the way of negotiating the Treaty were domestic party issues in the United States "plus the tense strain between London and Washington which had developed over Palestine". Wrong went on to say, "When Lovett appeared before the Senate Foreign Relations Committee in closed session to support the Vandenberg resolution, the Palestine issue was injected. Rusk the other day mentioned the personal message which he had sent to you by me a few weeks ago to the effect that an understanding over Palestine was necessary if the security pact was to be concluded, and asked me to tell you that he was still strongly of that opinion."

If differences of opinion between the United States and Britain over Palestine might have endangered the success of the negotiations on the North Atlantic Treaty, what would have been the effect of failure by Britain to grant independence to India? If independence had not been conceded in 1947, the newspapers in the United States would have been full in 1947 and 1948 of stories of mass civil disobedience in India, of the imprisonment of Mahatma Gandhi, Nehru and a host of their colleagues, of widespread disorders resulting in British troops firing on mobs. The result would surely have been an upsurge of anti-British feeling in the United States so great as to make the conclusion of a treaty impossible. Alastair Buchan, in his article in *Foreign Affairs* in July 1976, declared that "if Clement Attlee had not unilaterally defused the Indian issue by granting independence in 1947, it would have been difficult to sustain any intimate Anglo-American relationship in the post-war era."

Lovett told the British on 7 February 1948 that Bevin's proposal that the United States "enter with Britain into a general commitment to go to war with an aggressor" was premature. Before the United States could consider the part it might appropriately play in support of the Western European Union proposed by Bevin, there would have to be evidence of West European unity with a firm determination to effect an arrangement under which the various European countries were prepared to act in concert to defend themselves. Moreover, the United States Government did not have any very clear picture of exactly what Bevin's proposals for a Western Union really were, certainly not enough information to enable it to determine how it could best be helpful. "You are, in effect, asking us to pour concrete before we see the blueprints," Lovett said.

Bevin complained that "without assurance of security, which can only be given with some degree of American participation, the British Government is unlikely to be successful in making the Western Union a going concern . . . But . . . until this is done, the United States Government . . . does not feel able to discuss participation." The British considered that the United States had set them an impossible task. The United States considered that the British were making a "slanting effort to entangle [them] in European quasi-military alliances of agreements" — which was, of course, just what the British were trying to do. Here was what Bevin in his message of 6 February called a vicious circle.

The argument about how to get out of this vicious circle might have gone on interminably, and the result might have been no Treaty of Brussels and no North Atlantic Treaty. Indeed, some time in February 1948, before the Soviet coup in Czechoslovakia, the British Government had sent Gladwyn Jebb to Washington to sound out the United States on how far it might be prepared to go to support a West European defensive pact, and had recalled him before he spoke to the Americans "since the moment was deemed unpropitious". Even the shock of the Communist coup in Czechoslovakia on 25 February and the death of Jan Masaryk were not enough to budge the United States administration. Then on 8 March Halvard Lange, the Norwegian Foreign Minister, informed the British and United States Ambassadors to Norway that the Norwegian Government feared that Norway might face Soviet demands for a pact as soon as, or even before, the Soviet-Finnish pact was concluded. The Norwegian Government understood that the Soviet Union was confident that the result of the pact would be that not only Norway but also Sweden would do whatever the Soviet Union wanted. Lange asked what help Norway could expect if attacked.

The message from Norway was a catalyst. What had been unpropitious became urgent. Within two days of receiving the message from Norway — time for Gladwyn Jebb to draft telegrams and for Bevin and Attlee to approve them — the British proposed to the United States and Canada that officials of the three countries should meet in Washington without delay to explore the possibility of establishing a regional Atlantic approaches pact of mutual assistance under Article 51 of the UN Charter. All the countries directly threatened by a Soviet move on the Atlantic could participate: Norway, Denmark, Iceland, Ireland, France, Portugal, Britain, the United States, Canada and Spain once it had a democratic regime. On 11 March, Canada accepted Bevin's proposal; the next day the United States did so. Marshall demonstrated his sense of the urgency of action by proposing that the discussions begin early the next week, that is about 16 or 17 March — they began on 22 March.

It seems scarcely conceivable that the State Department would have become converted to the idea of a North Atlantic Treaty if it had not contained within its ranks two creative officers, Hickerson and Achilles, who fought for the Treaty from the end of 1947 against the doubts, hesitations and outright opposition of their seniors in the Department. When Hickerson was riding back with Acheson from the ceremony at which the Treaty was signed, Acheson said: "Well, Jack, I think this treaty is going to work. If it works, for generations there will be arguments in the United States as to who, more than anybody else, is responsible for it, but if it doesn't work, there will be no damn doubt, you did it."

Once the State Department had been converted to the idea of a North Atlantic treaty, it had the zeal of a convert and, like many new converts, it perhaps went to extremes in its efforts to propagate its new faith. Thus during the course of the negotiations, it put immense pressure on France, Belgium and the Nordic countries, backed up by the threat — explicit or implicit — that if they did not join the alliance the United States would provide them with little or none of the military equipment they urgently needed to strengthen their armed forces.

But mere zeal, even when backed by threats, would not have enabled a weak State Department to win over reluctant West European countries and a reluctant Senate. Fortunately for the alliance, the State Department possessed during the negotiations remarkable influence and power. For the first ten of the twelve months of negotiations, Marshall was Secretary of State and Lovett was Under-Secretary. Dean Rusk stated in 1976 that Lovett was Marshall's *alter ego* and that the "combination of Marshall and Lovett at the leadership of the Department of State has never been equalled in our history and is not likely to be again."

Lovett and Vandenberg were devoted friends and were accustomed to working closely together. It was Lovett who persuaded Vandenberg to support the idea of a treaty and to put the Vandenberg resolution through the Senate. Had it not been for their friendship, the difficulties of this task would have been formidable, perhaps insuperable. Marshall and Dean Acheson (who was in charge of the negotiations during the final two months, February and March 1949), had the confidence of President Truman. Truman knew how to bully or cajole reluctant senators.

When, in February 1949, Acheson's advisers were prepared to give in to the demands of senators to emasculate Article 5 of the Treaty (the pledge), Truman refused to give in. He said that only if it were "absolutely necessary after a stout fight" should the Administration agree to omit the reference in Article 5 to the possible use of armed force; he himself would speak to Connally. On 18 February, the day after he had spoken to Connally, Acheson secured the tentative agreement of the Senate Foreign

Relations Committee to the language of the pledge agreed to by the seven-power negotiating committee on 24 December 1948, with one change: the words "such action as it deems necessary, including the use of armed force" would be substituted for "such military or other action . . . as may be necessary".

As a Canadian, I like to believe that one of the factors which resulted in the success of the negotiations on a North Atlantic Treaty was the full support which the Canadian Government gave to the idea of a treaty from the middle of March 1948 onwards, support expressed in public speeches and in confidential discussions with other governments. When the United States Administration hesitated to commit itself to the idea of a treaty, Canada, as well as Britain, put to the Administration the arguments for a treaty. When France and Belgium hesitated to commit themselves, Canada contended in lengthy messages to their Governments that their national interests would be served by the conclusion of a treaty.

How far Canadian arguments and Canada's support for the Treaty had any effect in Paris or Brussels, I do not know. I do know that they did affect the views of at least one officer of the State Department, and that a very important one.

Louis St Laurent was Foreign Minister of Canada or Prime Minister during the negotiation of the Treaty. From March to June 1948, when the United States Administration was still not converted, St Laurent argued the case for the Treaty in seven public speeches. He said at the time that he was conducting "a crusade . . . for the completion of a Western Union or North Atlantic regional pact". George Kennan's conversion was at least in part due to St Laurent. On 24 May Kennan told Marshall that St Laurent's speech of 29 April had added a "new and important element" to the problem of a North Atlantic security pact and that he no longer opposed it.

Is it possible to draw a moral from the story of the miraculous birth of the North Atlantic alliance? I suggest that there is a moral to be drawn — as Robert Lovett said, "If domestic politics were the art of the possible, international politics might be called the art of the almost impossible."*

My generation made the North Atlantic alliance as a partial

Foreign Relations of the United States. 1948, Vol. III, p. 229.

and imperfect response to some of the challenges which then confronted the Western world. Another generation must decide what new ways of thought, what new domestic and foreign policies, new international agreements and institutions are required to make a creative response to the challenges which now confront the world, challenges which can be met only by practising the art of the almost impossible.

FATE OF THE NORDIC OPTION

The absence of guarantees for a Scandinavian defence association turns Norway firmly towards the Atlantic Alliance

Grethe Vaernø

NATO is today a fact of life in Norway. Some 80 per cent of the population believe that the United States will come to the assistance of Norway in case of attack; 70 per cent hold the opinion that membership of NATO contributes to peace and security for the country.

The Norwegian road to the Atlantic Alliance was, however, a bumpy one, even if by astute political leadership the Government landed Norway firmly in the Western camp in time to sign as a founder-member on 4 April 1949.

The only serious opposition to the Alliance was found within the ruling Labour Party. The political fight left visible scars, and every initiative that may be construed as leading to an escalation of tension or a military build-up still causes trouble for the Party and leads to the same kind of divisions.

This situation lends more than academic interest to the history of Norwegian entry into the Alliance. The debate never really ended. The "Nordic solution" remains as a mirage for the opponents, and the debate erupts time and again whenever new documents are released. Norwegian political memoirs are, however, singularly uninteresting. On thorny issues the traditional statements are reiterated, leaving the impression that the story was never fully told. The political system does not lend itself to revelations, and Norwegian archives remain closed to general scrutiny while the American and British archives of the period are now open to the public. The frustrated investigative efforts have kept alive the suspicion that there was much more to the story than ever met the eye. This theory is very welcome indeed for the critics of Norwegian membership of NATO, who are relatively few in number but highly vocal and articulate and capable of exerting influence within the Labour Party.

After thirty-five years, the most important considerations from 1948 are still with us: the "special relationship" with

Sweden; the wish to maintain an inoffensive, low profile toward the neighbouring Soviet Union; the question of whether the West European or the Atlantic ties are the most important to Norwegian security — and the importance of Norway to Western and American defence.

The close kinship and intertwined history of the Nordic area have, naturally enough, a profound effect on the foreign policy of Norway. In 1948 there was a strong desire to reconfirm the links, especially with the brother social democratic government of Sweden, that had been put under a severe test during the war. It is a paradox, however, that since the dissolution of the Union with Sweden in 1905, Norway has turned *westward* whenever a choice has arisen.

The Swedish relationship is, anyway, as much a reality today as thirty-five years ago. The "Nordic Balance" — the interlinked security position of the five Nordic countries — still exercises, at least in theory, a moderating influence on both countries. The state of Swedish defence, its ability to ward off violations of its neutrality and thus act as a buffer for the defence of Norway, remain crucial. Swedish-type neutrality still lingers on as a security policy alternative for those Norwegians who feel that NATO membership exposes Norway too strongly to the threat of being a pawn in the global power struggle between the two Super-powers.

It is not surprising, therefore, that the critics of Halvard Lange, Foreign Minister at the time Norway became a member of NATO, have asked whether he really did his utmost to find a Nordic solution. There were of course attempts to find alternatives to full Norwegian participation in a Western pact. There was no happiness at the thought of departing from the idea of "bridge-building" after the war. Socialist nostalgia, the respectability of the Communist Party after wartime underground cooperation, and quite relaxed attitudes toward the Soviet Union following its uncomplicated withdrawal from the invaded parts of the country's northernmost province, Finnmark, in 1945 all played their part.

Merging with these underlying impressions were fears of presenting too much of a challenge to the Soviet Union, of increasing tension and adding to, rather than eliminating, whatever danger there might be to Norway in an encirclement of the

Soviet Union. In addition, possible adverse consequences for Finland were part of the equation.

As a result, Norway inquired quite pointedly about possible demands from the prospective Alliance that she should cede base rights to the Western powers. The first answer implied that the United States did not require facilities in Norway, which was quite correct at that time. On the second occasion, Norway was informed that this was not a matter for the United States to decide: the requirements of the Alliance would be discussed by the Alliance itself. With this in mind, the Norwegian unilateral declaration that Norway would not join any treaty which would oblige it to allow foreign bases on its territory can be seen not only as a reassuring statement to the Soviet Union, but also as a pre-emptive move *vis-à-vis* the would-be Allies.

This rather ingenious declaration is supposed to give Norway considerable freedom of action and more options in times of tension. Some observers feel, however, that the way the policy has been put into practice has contributed to a certain rigidity in Norwegian policy, when even minor adaptations to changed external conditions become difficult to implement without giving a pretext for accusations of offensive actions toward the Soviet Union.

The mainstream of Norwegian sentiments, however, was strongly pro-West — particularly pro-Great Britain and the USA. Many factors had left their mark: the failure of Norwegian neutrality; the experience gained by five years of war; the habits of cooperation, and the embryonic plans from 1942 onwards for a post-war Atlantic alliance; the tremendous importance of the USA in European defence; and the increasing antagonism between the Soviet Union and the West.

In a rather quick response to the Bevin speech of 22 January 1948 calling for Western defence cooperation, Sweden put forward the idea of Nordic defence arrangements. This move was seen primarily as an initiative to give Norway and Denmark some option other than joining a Western pact which might increase Soviet pressure upon Finland and raise the risk of Western bases being established on Norwegian soil — too close to Sweden for comfort.

Discussions began during the summer of 1948 and lasted till the end of January 1949. To most observers, the standpoints of the partners were irreconcilable. Denmark, for its part, would have gone along with any Nordic solution either non-aligned or attached to the West. Sweden, on the other hand, could not stretch its neutrality beyond a non-aligned Nordic Defence Union and could not accept that any of the member-countries of such a body could have ties to a third party. It accepted, apparently, that this Union would be in need of arms supplies from the Western countries, and seemed willing to gamble that they would get such supplies once the Union was a fact — but not as long as the USA thought that Norway could become a fully fledged member of the NATO Alliance, bringing Denmark with Greenland along as well.

On the Norwegian side, the optimal hope was that it would be possible to bring all the Nordic countries together into some kind of cooperation with the Western Pact, thus assuring arms supplies on "lenient terms", a guarantee of assistance in case of attack, and staff talks to ensure effective cooperation in case of war.

It is quite clear that Lange and most of the other foreign policy spokesmen in touch with the US Embassy had no faith in the Scandinavian solution simply because they did not believe that Sweden would be willing to allow the kind of cooperation or understanding with the Western Powers which they thought indispensable for deterrence as well as effective defence.

Nor did they believe that a *neutral* Nordic defence arrangement would receive any such preferential, unilateral benefits from the West, either in the form of arms or of guarantees. A crucial question therefore is whether their disbelief was conditioned by their own dislike of the whole Nordic, neutral idea, or even whether they actually did not explore this possibility in good faith but, on the contrary, collaborated with the Americans to prevent it.

For such preferential treatment from the West to be probable, however, Norwegian territory would have to be credited with a strategic value that would warrant a privilege beyond anything accorded to other European nations. Norway's main strategic value, however, lay in denying it (including Svalbard, or Spitsbergen) to the enemy. The Americans did not foresee the military and technological developments that have subsequently increased the importance of Norway as a "listening post" in the

North, and as a support area in the struggle for control in the North Atlantic as the Soviet Northern Fleet assumes global dimensions, constituting a real threat to the lifelines across the Atlantic.

However, the documents point out that the United States was well aware of the importance of Greenland and Iceland as stepping stones in the defence of Europe, and the political and psychological impact of Norwegian and Danish membership. Besides, the dislike of neutrality was thoroughly entrenched in the political atmosphere of the post-war years.

In theory Norway, Denmark and Sweden had eased their positions in the Nordic talks. Even as late as 8 February 1949, when the talks had actually broken down, Lange said in Washington that "if the US and Britain believed the Scandinavian pact is the best solution for the three countries, their three drafts would be reduced to one in a couple of days." The Norwegian minimum demand included an advance promise of American arms deliveries on favourable terms (entailing staff talks, to which the Swedes subsequently objected) and Western goodwill. The foreign policy leadership in Washington which, besides Dean Acheson as Secretary of State, included some less hawkish men, took a quick second look and decided against a Scandinavian pact.

It is plausible to argue today that the Americans consciously used the denial of arms supplies on favourable terms in order to undermine a neutral Nordic arrangement that would have prevented Norway and Denmark from joining NATO. Lange, for his part, was probably not interested in any indications that such supplies might have been forthcoming, mainly because arms would not have been sufficient as a deterrent, and secondly because that would only have strengthened the Swedish negotiating position while Norway still hoped to find an arrangement which would have involved Sweden somehow in the defence of the West.

However, Lange repeatedly asked for and received reconfirmation of the American view, already expressed in September 1948, that those countries that joined the common Atlantic effort would be served first out of very scant supplies. But in the light of later developments, as well as the exchange of messages between the State Department and the Nordic capitals, there can be little doubt that the "arms argument" was a very handy and "objective" instrument that offered itself at almost no cost.

What a Nordic arrangement would really have needed was some kind of unilateral guarantee from the United States indicating that the defence of Norway would be vital to American national security and to Western defence. But even if such a guarantee was indeed accorded to some groups, it seemed — objectively — out of the question for Norway when the alternative was Norwegian membership of the Atlantic Pact. On the contrary, when Lange inquired, in an atmosphere of crisis following the Czechoslovak coup and rumours of a pending "offer" of a non-aggression pact with the Soviet Union in the beginning of March 1948, whether Norway could expect assistance from the wartime Allies in case of attack — the answer was a friendly but bland statement that the USA was considering the security of the whole area.

Everybody, at this time, wanted guarantees from the United States, but the State Department emphatically stated and restated that the United States did not contemplate giving guarantees to anyone; what was under discussion was a collective defence arrangement. The general political trend was also running against one-sided, open-ended commitments. With hindsight, one might perhaps conclude that it would have been in the United States' interest to help the Scandinavians stay out of the Soviet sphere of influence by all means possible — but in 1948 this was not so apparent.

Almost from the moment of Bevin's speech, high-ranking Norwegians showed keen interest in a Western security arrangement. Very enthusiastic envoys in Oslo and Washington speeded messages across the Atlantic conveying the impression that talks and negotiations about a Nordic option were more like obligatory exercises than political realities; that there was deep anxiety in Norway at the security situation; and that Norway looked to the United States for the solution to these problems.

The Norwegian initiatives were probably necessary in order to stimulate an active American commitment to Norwegian membership. The inclusion of Norway was not a foregone conclusion, even if from the very first background papers of early 1948, prepared by the State Department's Policy Planning Staff, Norway was included as a matter of course among the many states that would need some security arrangement. The problem seemed to be to find the kind of arrangement that would, on the one hand, foster an integrated, political cooperation in Western Europe, perhaps by a bilateral agreement with the rather limited Brussels Pact of 17 March, and on the other, allow for membership of states that would not join the Brussels Pact. While European integration was a prime target for the United States, Bevin seems to have concentrated on finding some way to entice the United States to commit itself to the defence of Europe. Bevin, therefore, happily grabbed the chance offered by the Norwegian inquiry of early March and pointed to the immediate threat to Norway, urging that negotiations be started on the wider concept of a collective security arrangement that could comprise all the countries exposed to Soviet aggression.

When, after the preliminary talks between the United States, Canada and Great Britain, these countries met with the full group of the Brussels Pact for the Washington exploratory talks in the early summer of 1948, the United States had, for its part, decided on the desirability of Norwegian membership as the key to the Nordic area. The United States and Canada flatly declared at one point that without the participation of the "stepping stones" (Greenland and Iceland) they were not interested in an agreement only with members of the Brussels Pact.

France and the Benelux countries reacted with surprise and some opposition. They were annoyed by American stalling, caused partly by domestic politics and partly by the wish in Washington to see the Europeans go ahead with the consolidation of their own cooperation. But the Europeans did not want to discuss enlargement of the European group until they felt that the United States was really committed to the idea of an alliance.

By the beginning of September, agreement was reached on a collective arrangement between individual countries, and Norway received a confidential enquiry from Washington on whether it wished an invitation to join further talks. Referring to the ongoing discussions with Sweden, it asked for a postponement. What Norway was not told, however, was that the agreement allowed for the possibility of "graded membership". At this point, the State Department staked considerable prestige on the Scandinavian issue, and launched a policy of applying pressure on Sweden "to let Norway alone", as one participant forcefully expressed himself, and on Norway a policy of benevolent, "disinterested" concern for security and wellbeing.

When, after the failure of the Nordic talks and with increased pressure from the Soviet Union early in 1949, Norway requested an invitation to join the final negotiations, France raised objections and tried to trade Norwegian membership for the inclusion of Algeria, but this was rejected by Dean Acheson — thus the road was finally cleared.

It is futile to speculate on what might have been if Norway had decided on a different course. We shall have to content ourselves with the same ambiguity that surrounded the decision-making process at that time, without succumbing to the common urge of posterity to simplify the past as a response to the confusion of the present.

DIPLOMACY ON THE SPOT
The role of the Italian Embassy in Washington in the negotiations leading to the North Atlantic Treaty

Egidio Ortona

More than thirty-five years after its creation, the Atlantic Alliance still constitutes the most viable institution for the defence of Western democracies, and Italy has loyally, steadfastly and conspicuously contributed to it men, equipment and ideas. It is therefore instructive to look back at the circumstances under which Italy entered the Alliance, and the obstacles which it had to overcome, on a national as well as on an international level in its relations with its potential allies, before its admission. These events evolved in the relatively short span of a year, from April 1948 to April 1949. It was a period packed with intense and tightly interwoven developments, in which the Italian Embassy in Washington played a crucial role. It is from that perspective, taking advantage of my own notes taken as the situation unfolded while I was serving in the Embassy, that I shall try to reconstruct those events.

I shall outline, as a basic premise, the areas of difficulty that confronted the Embassy, and describe the often arduous battle it had to wage to pursue its goal, as it did under the relentless and dedicated direction of the then Ambassador in Washington, Alberto Tarchiani, to make Italy an equal partner with the other members of the Atlantic Alliance, with all the necessary consensus from all sides which would qualify the Italian contribution as a positive and valid one.

There were four areas of difficulty. First, in the internal political situation in Italy there were no clear signs of support for Italy's entry into the Alliance — at the outset anyway. In some sectors of the Left, calls for neutrality were in fact quite pronounced. Secondly, there was diffidence on the part of certain potential members of the Alliance with whom Italy had been at war a few years earlier, and with whom it had signed a peace treaty only a couple of years before. Thirdly, the Americans seemed to waver, being concerned on the one hand with possible dissensions with past and future Allies, and on the other hand fearing to antagonise a country such as Italy with its undeniable geopolitical importance and with large ethnic groups of Italian origin within its own territory. And fourthly, the definition of the Alliance itself created a dilemma because, as it referred solely to the North Atlantic, it gave obvious pre-eminence to the countries of Northern Europe. However, it was impossible to ignore the importance of the Mediterranean. As a result of the arguments which originated from these difficulties, the Italian Embassy not only became deeply involved in the problems, but was also required to promote and stimulate initiatives and actions, playing the role of an essential, active and constant adviser for its Government.

In the following account, we shall touch in rapid succession upon the variety of intricate events which were unleashed by these difficulties. It should be said at the beginning that Italy's participation in the Atlantic Alliance finds its roots in the establishment of the Western Union at Brussels in March 1948 involving five European countries — Britain, France, Belgium, the Netherlands and Luxembourg. To this group of countries, the American Government agreed in April 1948 to supply military aid. Already in a private conversation with Ambassador Tarchiani on 27 April, the US Assistant Secretary of State John D. Hickerson let it be known that Italy's presence in that group was considered by the US Government a necessary element and was nearly a foregone conclusion. Such a statement by Hickerson, repeated on various other occasions, and in particular during another conversation on 10 July, became a strong catalyst for action by Ambassador Tarchiani. During a visit to Rome in May, he had already argued in favour of Italy joining the Western Union, sensing that it represented the beginning of a treaty of vastly greater importance, to include Western Europe, the United States and Canada. It was in fact a Union that would develop, a few months later, into a true alliance.

In Rome the Prime Minister Alcide de Gasperi and the Minister of Foreign Affairs Carlo Sforza, while appearing convinced that Italy's destiny should not be dissociated from that of the other countries of Western Europe, harboured fears that Italy's formal participation in the Western Union would encourage calls from the Left for neutrality. This would alienate the Americans, who now appeared "firm in their conviction that the Italian borders should be defended with the same decisiveness and energy as was promised for the defence of the French borders" (as stated by the US Ambassador in Rome, James Dunn). These perplexities were enhanced by the surfacing in certain political circles of strong resentment and criticism of de Gasperi who, with one bold stroke, had just excluded the Communists from his government. This was a daring decision to take, especially as it occurred after his visit to the United States some months before.

Tarchiani's talks in Rome with a number of politicians and high-ranking officials of the Ministry of Foreign Affairs in May 1948 were meant to bring to the surface the growing need to take a position, though not too hastily, given the uncertain internal atmosphere in Italy with regard to participation in the Western Union. Tarchiani finally succeeded in obtaining from the leading Italian politicians and diplomats agreement that the Americans should be assured that Italy certainly did not want to dissociate itself from them, particularly at a time when American aid was giving a vital boost to the country's economic recovery. From de Gasperi and Sforza, Tarchiani also obtained consent to tell the American Government that Italy favoured American initiatives to study military plans for the defence of Italian borders. Beyond that they would not go, still adhering to the tactics of avoiding any statement regarding Italy's entry into the Brussels Union.

It has to be remembered that the internal Italian situation was soon to be further complicated by incidents such as the attempt in the following July on the life of Communist Party Secretary Palmiro Togliatti. A speech given by Sforza in Perugia on 7 July 1948, containing phrases which implied an Italian position of equidistance between the two blocs, led the Italian Embassy in Washington to send a series of strong messages to Rome, the more so because of the progress which was underway within the Western Union. On 27 July Tarchiani sent a strong letter to Sforza again urging him in rather blunt terms seriously to consider the possibility of Italy joining the Western Union. He reinforced his warnings with the backing of other Italian embassies (primarily from Ambassador Pietro Quaroni in Paris).

He stressed that it was essential to counteract the opinions which were being expressed by Governments in the Western Union according to which Italian public opinion was not yet ready to adhere to the Union. Such initiatives by Tarchiani were all the more courageous since the State Department itself (particularly Hickerson) maintained a policy of caution. This was due partly to the awareness of the Italian Government's difficulties because of the still lingering doubts of Italian public opinion, and partly to the belief that there was no need to urge the Italians into action, as any serious discussion about military commitments would have to await the meetings of the Senate Foreign Relations Committee at the end of the year.

It was during this period, when Tarchiani and Italy's other ambassadors kept spurring on Sforza (who continued to reply that Italian public opinion had not yet "decided"), that an official of the Italian Desk of the State Department, Bill Sales, while discussing the Italian position toward the Western Union, made the lapidary comment, "Stop this nonsense of talking about neutrality." This comment, which stood in some contrast to Hickerson's cautious and measured style, had been partly inspired, according to Sales, by certain reports reaching the Department through military channels, which had underlined the growing suspicion among the British and French General Staffs regarding the still uncertain attitude of the Italian authorities.

I have mentioned that we were encouraged by our fellow-embassies, particularly in Paris, to urge our Government to join the Western Union. On a visit to Europe in the summer of 1948 I stopped in Paris, and was able to verify Ambassador Quaroni's determination in pressuring our Government. He was pleased by my news of the exchange of messages between Washington and Rome on that disturbing theme, and did not hesitate to criticize our "equivocal and non-existent foreign policy", saying that "Italian politicians were full of fear" and urging us to persevere in our campaign for entry into the Western Union, since it was an essential part of our foreign policy from which we could not withdraw.

In Rome, where I went after Paris, I found the Ministry in an unreceptive mood to Ambassador Tarchiani's suggestions; this was because, although there was potentially a favourable disposition towards Italy joining the Union of the Western powers, it was clear that it could not take place without some compensation which would satisfy public opinion. The latter still felt the heavy weight of the peace treaty and was much more inclined to favour flexible formulae. Hence the intention was to put off any public discussion of the subject. However, it was pointed out to me that in spite of the wavering of the Italian Government, the Americans did not need to have any doubts, because if members of our Government took any position not in line with American policies, this would have brought about their downfall. The United States had to be patient and allow the Italian politicians gradually to convince their public opinion.

During my short visit to Rome, I was received by the Foreign Minister. Sforza began by asking me point blank, "Why do Tarchiani's statements on the Western Union not correspond to what Ambassador Dunn tells me in Rome?" I pointed out that it was partly due to the fact that the Americans were aware of the danger of pushing our Government too hard for a public commitment to join the Union, which was not yet unanimously acceptable, and partly due to the lack of enthusiasm of some of our allies, particularly the British. In any case, Sforza agreed that it was madness to talk of equidistance and neutrality. On the other hand, he felt that the Americans should have more faith in Italian politicians, and let time take its course.

On his return from leave in Italy, Tarchiani contacted the State Department, where Hickerson expressed his opinion, this time more clearly, pointing out that the British and French were referring to the hesitancy of the Italian Government: this made the Italian decision to join the Union more essential and urgent, with all its implications, including the readiness to take part in the common effort.

A new element entered into the dialogue that had developed between the Italian embassy in Washington and the State Department. In early October the State Department sent us a confidential message that the Chief of the General Staff, Omar Bradley, wanted to invite his Italian counterpart, General Efisio Marras, to Washington. The purpose of this move was to strengthen the Italian armed forces so as to make them reach a degree of efficiency which would be in line with the scope and needs of the Western group and, according to a message from

General Bradley to Tarchiani, enable them to become a part of the Western defence programme.

Because of these aims, needless to say, the task of arranging such a visit became rather complex, as uncertainties and doubts kept reaching us from Rome about the advisability of proceeding with it, at least for the moment. In late October, the Embassy received a telegram with instructions to delay the visit as well as any announcement of it; but the Italian Desk of the State Department obviously expressed its strong disapproval and made clear allusions to the risks if the visit did not take place. The decision to delay the visit was a reflection of the still undecided state of Italian public opinion, and resulted in repeated expressions of worry and regret by our friends at the State Department who, at that point, informed us unequivocally that they intended to keep pushing for Italy's entry into the Western Union. They had actually gone a step further by sending instructions to the assembled representatives of the five member countries of the Union in Paris, to the effect that the United States definitely favoured Italy's participation.

Our Washington Embassy faithfully communicated all these pressures to Rome, in such a way as to induce the Italian Government to proceed with General Marras' visit. This finally took place in early December, fulfilling the hopes and expectations of both the State Department and our Embassy. The visit had a special impact, not only because of the efficiency and authority of our Chief of General Staff, but also because it implicitly caused Italian public opinion and politicians to face the inevitability of thinking along military lines, the more so because the Brussels Treaty was then gradually evolving as a prelude to the formation of an "Atlantic Union" with all the characteristics of a military alliance.

The visit of General Marras greatly accelerated the movement towards the desired goal. It included a successful meeting with President Truman, which prompted a journalist friend to comment that neither the French nor the British had had such a reception.

In the mean time we were receiving information, especially from the Ambassadors of Luxembourg and France, that at the meetings of the five member-countries of the Western Union there was continued opposition to our entry into the Pact from the British and little enthusiasm for it from the Benelux countries. This led Tarchiani to warn Rome that we were running the risk of the five countries adopting a formula whereby "participation in the Atlantic Pact would be reserved to those countries facing the Atlantic ocean, plus those in the Brussels Pact." This simply meant that it was becoming ever more urgent for Italy to declare itself.

The year 1949 began with a positive development, giving us reason to believe that the Washington Embassy's efforts were beginning to have an impact. In fact, on 4 January we received a telegram signed by Foreign Minister Carlo Sforza (we described it as "acrobatic") which authorised us to confirm officially to the Americans that we were ready to "come to an agreement with them in the interest both of Italian independence and of European unity". In the same telegram we were informed that a memorandum would follow that would fully "clarify" our policy.

As we waited for the memorandum with almost morbid curiosity, anticipating that its purpose would be to allay the doubts of Italian public opinion, we noticed with some displeasure an article written by the well-known political commentator Walter Lippman, in which he discussed the Atlantic Pact and spoke of the expediency of excluding Italy. In the following weeks he repeated this theme, stressing the need to restrict the geographic range of the pact and to create a belt of neutral "buffer states", including Italy, around the members of the future Alliance. This position did not make matters any easier for us at a time when we were involved in a series of diplomatic manœuvres in the United States and elsewhere, loaded with difficulties and uncertainties. (I had the occasion to see Walter Lippman at a dinner during those crucial days, and did not miss the opportunity to tell him how disappointed and hurt we were by his attitude. He explained his conviction that it would not have been wise for the US government to encourage the participation in the Union of countries with a difficult internal situation, particularly when Americans themselves were not in total agreement, as there was still the possibility that Congress would reject such a binding commitment as an Atlantic Pact. Furthermore, he added that he was discharging a duty, for writers also had a responsibility, and that nothing would have been worse for the United States than to make commitments with recalcitrant and politically unprepared European countries.)

Between the telegram and the announced explanatory memorandum, another telegram was sent from Sforza to our Ambassador Tarchiani in Washington, classified as "for your eyes only", in which the Ambassador was instructed to explain that the content of the forthcoming memorandum was to be prefaced with the qualification that "your and my personal opinion is that if we were called upon to participate in international discussions this would not create embarrassment of an internal political nature." This additional message from Sforza made the wait for the memorandum all the more baffling as it forewarned us of rather intricate explanations. And in fact, when the memorandum finally arrived, we noted that there was more division than agreement over our adherence to an Atlantic Pact. The American Government was requested to give a prior pledge of intervention in the case of agression against Italy, this being a condition for our participation. It confirmed the existence of political forces in Italy that were against the Brussels Pact, and doubtful over our eventual entry into an Atlantic Pact. Only when the Americans confirmed this guarantee would Italy officially request participation. Together with the memorandum came a letter from the Secretary-General of the Ministry of Foreign Affairs, Vittorio Zoppi, in which, while he admitted that the memorandum was not satisfactory, we were nevertheless asked to refrain from arguing with the Ministry given the internal problems with which it was confronted: the document had to be interpreted as a "basis for developing negotiations". Tarchiani replied with a telegram saying that, barring any instructions to the contrary, he would forward the memorandum, expressing the Italian Government's position in the following way: "Italy has a difficult internal situation. The Government, however, is in a position to join the Pact and to have it approved by Parliament provided all the necessary prior guarantees are given, including their extension to the Free Territory of Trieste."*

---

*Provisions for the Free Territory of Trieste were contained in annexes to the Peace Treaty between the Allies and Italy. At this time, the area was divided into two zones, 'A' under Anglo-US occupation and 'B' under Yugoslavia.

As for the procedure to follow in the presentation of the memorandum, we considered different possibilities: forwarding the entire memorandum as received, or forwarding it with a letter of explanation, or, finally, forwarding it with another memorandum from the Embassy itself explaining the details more explicitly and exhaustively, on the basis of a series of previous instructions from Rome to us and other Embassies (one example being the unequivocal phrase used by Sforza in a message to Ambassador Pietro Quaroni in Paris: "We intend to join the Alliance"). The third method was chosen, and on 12 January 1949 Tarchiani proceeded hurriedly to the presentation of the documents in an official meeting with Hickerson, the US Assistant Secretary of State knowing that time was precious because of the State Department's need to study the memorandum before a planned meeting of the ambassadors of the five member-countries of the Western Union. In the same meeting with Tarchiani, Hickerson posed a few questions, in particular whether we would have any objection to our simultaneously joining the Brussels Pact and the Atlantic Alliance. As foreseen, a few days later, on 15 January, the meeting of the five ambassadors took place with Under-Secretary of State Robert Lovett. Since the Americans informed the group of the Italian intention to participate as indicated to them three days earlier, the question of Italy's membership became the main and almost exclusive issue discussed at the meeting.

We were later told by the French Ambassador Henri Bonnet (always extremely friendly and favourable to our admission) that the Americans had not been forceful enough in supporting our cause, especially in view of the insistent objections of the British, who incidentally had also aired the hypothesis that the extension of the Atlantic Pact to Italy would endanger the chances of the entry of the Scandinavian countries because of the fear that the adherence of Italy could shift the centre of the Alliance toward the Mediterranean.

While the Italian Government was slowly strengthening its conviction that its intention to join had to be more clearly expressed, objections by other countries were multiplying. Difficulties came from the British (according to a telegram from London, Bevin said to our Ambassador there, Ludovico Gallarati Scotti, that it could be "in your own interest to delay your entry"), as well as from the Canadians, the Dutch and the

Scandinavians. In fact, when the Norwegian Foreign Minister, Halvard Lange, came to Washington virtually to give his country's agreement to joining the Alliance, he also expressed Norway's opposition to Italy's entry. These were not the only instances of opposition, however, as others, derived from US internal reactions, were gradually to be added. In fact, as the diplomatic negotiations to create the Atlantic Pact proceeded, the problem could not but become a matter of study in the machinery of Congress, where the Senate, having recently made considerable efforts to accept the Marshall Plan, was trying to limit further military and financial commitments by the United States; thus it favoured a Pact with as limited a geographical area as possible. To counteract these tendencies, a plan to defend our position developed among the friendly sections of the State Department. In fact Hickerson, at a meeting on 18 January with the representatives of the five Western Union countries, spoke firmly of the American Government's desire to see Italy enter the Atlantic Pact. But among the five it was only Henri Bonnet who pleaded our cause with authority, courage and conviction, mainly to prevent Italy from being relegated in the Pact to a category of second-rate countries. It was shortly before the end of January that our entry began to appear likely: on the 29th we learned from the State Department of the formal British announcement that if the Americans wished it, the British Government would pronounce itself in favour of Italy's entry into the Atlantic Alliance.

The battle then shifted sharply towards the American Congressional scene, where doubts and apprehensions were growing as a result, among other things, of heavy pressures for guarantees and commitments by the Norwegians, and of manœuvres by other countries. Because of this, the decision-making process within the US Congress was slowed down. In addition, we were informed that the President of the Senate Foreign Relations Committee, Senator Arthur Vandenberg, had, in a private conversation, expressed the opinion that a requirement for entering the Atlantic Pact would be membership of the United Nations Organisation, a status which Italy did not enjoy at that time. The new Secretary of State, Dean Acheson, for his part had taken it upon himself to become the spokesman for the group of the seven countries (the five of the Brussels Pact plus the United States and Canada) whose many objections and uncertainties were spreading to the American Congress. One felt that the debate on the Pact was coming to a halt, and that, in the consultations between Acheson and the congressional leaders, the traditional isolationist caution of the Americans was still to be reckoned with, especially because of the opposition within the Senate to definite military commitments.

After various requests for a meeting, Tarchiani finally saw Acheson on 17 February. It was not at all an encouraging meeting, Acheson having stated at the outset that the problem of the Atlantic Alliance, both in general terms as well as with regard to Italy's participation, was still under study. The Italian question, he said, could only be resolved with the agreement of the Europeans. Tarchiani asked whether the assurances given two weeks earlier by Hickerson not only to him but also to the Europeans as to America's desire to have Italy enter the Alliance, still held good. Acheson avoided the question, saying that he had only been in office for two weeks. "So should we deduce", Tarchiani interjected, "that your appointment may entail a change in policy?" "No", he replied. "but we have to coordinate everything with the Europeans."

This meeting had been very disappointing indeed. Swords had been crossed on both sides, and officials of the Italian Desk of the State Department felt that the discussion had not been fruitful and constructive, having even registered some harshness and incomprehension. However, it all served to show us that, despite the assurances we had been given, the situation was still far from being clear. It also revealed that the problem involved Congress, requiring the utmost attention and the employment on our part of the greatest efforts both in congressional circles and in the higher echelons of the State Department. In addition, we felt that, as we reached the final stages, we were approaching the study of clauses defining the "legal" commitments of the Pact. We were thus to concentrate particularly on the Legal Affairs section of the State Department, which was headed by Assistant Secretary Ernest Gross. A young up-and-coming official in that section was Walter Surrey, who was particularly sympathetic to Italy because, two years earlier, he had headed with success and friendliness the US delegation for the agreements with Italy relating to the financial settlement of the aftermath of the war. It was through Surrey that we sought to reach Gross and, in turn, the Secretary of State.

In Congress, we had substantial influence with John Lodge, who at the time was Congressman for Connecticut. Lodge was very actively interested in casting himself as champion of Italian interests (proof of this was his impassioned speech on the subject at the beginning of his term in 1947). He was motivated both by the conspicuous number of Italians in his constituency and by having an enterprising and dynamic Florentine wife, Lodge was particularly important to us as his brother Cabot was a member of the Senate Foreign Relations Committee. Cabot was very competent in problems of foreign policy and was a friend of the President of the Committee, Arthur Vandenberg. We were in growing need of friendly supporters, particularly at a time when the question of Norway's entry into the Alliance was gradually taking on greater weight. During a meeting of the six ambassadors with Acheson on 20 February, the Secretary of State had proposed that Norway be admitted immediately. French Ambassador Bonnet objected, saying that this should only be done simultaneously with Italy's acceptance. A heated discussion developed and Acheson severely admonished Bonnet, saying he was taking on a grave responsibility by delaying Norway's entry. Italy's acceptance, he added, could be decided only in a later meeting, and Norway's admission was also more urgent in relation to existing Soviet threats. The meeting adjourned having made no progress, its participants preferring to consult their governments. Our hopes seemed very slim that day since the defence of our cause apparently rested mainly with the French Ambassador. Nor had we been cheered by news from Paris that the French Government would not continue to support us for a *sine qua non* simultaneous admission with Norway. Instead it would demand that if Norway were accepted Italy should be admitted immediately afterwards, a situation certainly fraught with danger for us.

It was during the first week of March that a series of developments finally brought us near an unconditional admission. The starting-point was a telegram read to us over the telephone by Secretary-General Zoppi, who in the preceding days had not minced words with the Norwegian Ambassador over the obstacles his Government's position had created for us. Zoppi told us that in the telegram the Italian Government was formally asking to be invited to participate in the Atlantic Pact.

Tarchiani moved quickly with the French Ambassador, and

through the mediation of Walter Surrey with Ernest Gross. Gross told him that Secretary of State Acheson had consulted the previous day with Senators Connally, Vandenberg and George, who, according to Acheson, had objected to Italy's immediate entry. A meeting of the Ambassadors took place soon afterward at which it was decided that Norway would be accepted, and the question of Italy's entry would be delayed for two days, along with the entry of three newly added countries, Denmark, Iceland and Portugal.

This turn of events was decidedly not in Italy's favour, particularly since Acheson had given the impression of wanting to delay the question of Italy for some time to come. Throughout that day, we moved quickly to mobilise our friends, counteract our opponents and enlighten the indifferent. Tarchiani called on Senator Fulbright (who was flattered by the gesture), and later spoke with Senator Connally over lunch. The Minister of the Embassy, Mario Luciolli, made the round of the Embassies to inform them of our instructions and of our presentation of an official note to all the countries concerned. I myself drove our friend John Lodge to Congress, giving him all the useful information that his brother might exploit for our cause at the meeting of the Senate Foreign Relations Committee which was planned for shortly afterwards. While all this was happening in Washington, a conclusive phase of negotiations was unfolding in Europe where, in London, the following formula was agreed upon by the Europeans: Norway would begin to take part in the negotiations but would only initial the original text which would, in any case, mention Denmark, Italy, Portugal and Iceland as countries which would finally participate. All would afterwards sign together. Rome cabled us saying that such a formula was acceptable. At the Italian desk of the State Department, where our position appeared to be improving, we were advised not to make any announcement of acceptance, given that the American Government seemed increasingly intent on promoting conditions as much as possible in our favour.

In fact, the meeting of Ambassadors in Washington on 4 March opened with a statement by Acheson. He said that after having consulted with the President and Senators of the United States, he could announce that the United States favoured Italy's entry. There was no objection from the Europeans; in fact the

British Ambassador Sir Oliver Franks, who had been remarkably friendly to us in the final days, stated during that meeting that on the basis of the opinion of the USA, his Government withdrew the doubts it had expressed on previous occasions. The question of initialling the agreement came up at the meeting, and only at that point was the Norwegian Ambassador allowed into the chamber, because Bonnet, consistently friendly to us, had been opposed to Norway's presence in the meeting while Italy's membership was being discussed. Despite all this, while a definite decision was evidently near, Acheson stated that he would still have to consult with the Senate. (I learned later from Lodge that when the subject of Italy's membership was about to be examined at the meeting of the Foreign Relations Committee, his brother had passed Vandenberg a note encouraging him to conduct the session in a way that would allow our unconditional admission.)

Only several days later did we receive an official invitation, as it had required the formal approval of all the governments. But when everything seemed to have settled in our favour, an embarassing last-minute hitch developed. The Council of the Italian Social Democratic Party, headed by Giuseppe Saragat, voted 8–7 against our participation in the Atlantic Pact (a particularly unpredictable development as Tarchiani, in the course of various meetings held in Rome in May, had been greatly gratified by Saragat's expressing the view that on East-West relations we should side with the United States). The position of the Social Demoractic Party caused obvious irritation among the friendly sections of the State Department. This, however, was offset by conversations between Sforza and the American and British Ambassadors in Rome, in which he assured them that the Social Democrats' position would not change the Italian Government's course. On 8 March Hickerson officially presented Italy with an invitation to join the Atlantic Pact, adding that we were all united in the common defence. Exactly four years had elapsed since Tarchiani, then the official Representative of an enemy country, had presented his credentials to President Roosevelt. After Hickerson's announcement, Tarchiani was radiant, and with reason. The invitation to participate crowned the work he had carried out with tenacity, wisdom, good timing, enthusiasm and dedication during twelve months of intense action. That same day, he told me that common sense required that he should send de Gasperi a letter of resignation for "mission accomplished".

During the next few days, a flicker of controversy still arose in some rather harsh telegrams from Rome, complaining that the text of the Pact, as formally announced to us, did not contain a clause which was expected and on which the Government counted for the parliamentary debate. It was the clause referring to the automatic nature of reciprocal military intervention between members of the Alliance. Nevertheless, when Premier de Gasperi officially announced the news of our adherence to the Atlantic Pact, in parliament on 11 March 1949, he gave a lively and impassioned speech on the theme of peace and security, and appealing to our American friends for adequate assistance to Italy in the framework of the Marshall Plan. The debate was concluded in the Chamber on 18 March in favour of the Atlantic Alliance by 342 votes to 170. Thus was sealed an arduous year of the most intricate and dynamic diplomatic exchanges. The nations that took part are still bound by an Alliance that is even more necessary and unavoidable today than when it was created.

PAUL-HENRI SPAAK – FUTURE SECRETARY GENERAL
Belgium sees its hopes fulfilled

Baron Robert Rothschild

Since the days of the League of Delos in 477 B.C., when Athens gathered together most of the Greek city-states to repel the Persian invasion, the world has not known an Alliance that has lasted as long as the one embodied by the Atlantic Pact. Nor has there been, at any time, one set up with the sole purpose of maintaining peace rather than waging war. It is a remarkable story, and the historians of the future will certainly judge the politicians and diplomats of the post-war period as men of outstanding vision.

Among them was Paul-Henri Spaak of Belgium, a small country of only 10 million people, but in a strategically crucial position. The favourite battlefield of its powerful neighbours throughout the centuries, Belgium adopted from 1839 a policy of armed neutrality in the unrealistic hope of keeping out of further wars. Notwithstanding the fact that Imperial Germany, in 1914, had treated Belgium's neutrality — in her own words — like "a scrap of paper", Belgian statesmen returned to it in 1936 when the Western democracies, rather than opposing the Third Reich, advocated appeasement and détente — the two terms were given equal weight.

Spaak too, as Foreign Minister, had been taken in by the fallacies of neutrality, but the disasters of 1940 radically changed his attitude. As early as 1941, as a member of the Belgian Government-in-exile in London, he advocated a close association among the European nations to ensure "security and prosperity". By 1943 he had realised that his ideas could be applied only to the Western part of Europe. In several speeches, in his conversations with Eden, and in an official memorandum in 1944, he advocated a political, military and economic union under the leadership of Great Britain. However, the British Government was not interested in what the Permanent Secretary at the Foreign Office later described as "foolish and premature expedients". Nor was Bevin any more prepared than Eden to commit his country to a Continental policy: when Spaak, at the end of 1945, proposed a close alliance between Britain, France, Belgium, the Netherlands and Luxembourg, the new British Foreign Secretary answered that nothing could be done before the United Nations had come into being.

To the major powers the United Nations was still the main area of interest: it represented a universal forum where the Soviet Union could and would play a paramount role equal to that of the United States and Britain, a role that neither Washington nor London yet wished to deny. But Stalin would rapidly make them change their minds.

The Soviet Union had been alone among the principal belligerents to end the war by annexing large chunks of foreign territories: the Baltic states, a province of Finland, a sizeable part of Poland. The Russians had become all-powerful in Warsaw, Belgrade, Bucharest, Sofia and Tirana. But they were not satisfied with an empire extending from the Baltic to the Black Sea. All over the world, they were busily taking advantage of the post-war difficulties by encouraging subversion. In Greece a cruel civil war had torn the country apart. In the Far East the Soviets had occupied Manchuria and delivered vast amounts of Japanese war material to Chinese Communist armies fighting the legal government. In Iran and Turkey they were making trouble. In February 1948 the Prague coup had made Czechoslovakia into yet another satellite of the Kremlin.

As early as 1946 Winston Churchill, while in opposition, was the first to sound the alarm with his celebrated speech on the "iron curtain" at Fulton, Missouri. By September 1948, in another speech given in Paris at the UN Assembly, Spaak acquired immense popularity by telling Vyshinsky, the Soviet representative, forcefully how much "fear" the expansionist policies of his country inspired in the Free World. He thus accurately expressed the mood of the West. Everybody involved in politics at that time remembers how Stalin had, by then, replaced Hitler in the minds of the people as the main force of evil.

The turning-point in the policies of Washington, London and Paris was probably reached as a result of the Kremlin reaction to the Marshall Plan (June 1947). It is often forgotten that Marshall Aid had been proposed to *all* European nations, including the USSR. Moscow's reaction was so negative, and the interventions of Molotov and Vyshinsky so deliberately hostile and insolent, that Bevin and Bidault finally concluded that all roads leading to cooperation with the Soviet Union were definitely closed.

The British Government, which had long been loath to give up the wartime constellation and commit itself on the Continent of Europe, was now the first to react effectively. When at the beginning of 1948 Norway seemed, in its turn, to be threatened by the USSR, Ernest Bevin in a historic speech in the House of Commons on 22 January expressed his deep concern at the impossibility of cooperating peacefully with the Soviets. After recalling all their misdeeds in the world, after describing how they had betrayed all their Yalta commitments, he concluded with sorrow that the time had come once again when the values of Western civilisation had to be defended against the threats of tyranny. "We have", he said, "to consider Western Europe as one unit — including Germany."

This was the outline of the policy which for years Spaak had been pressing the British Government to adopt. The very next day, we published in Brussels an enthusiastic communiqué stating our deep interest in the British Foreign Secretary's ideas and announcing that we were immediately going to consult The Hague and Luxembourg. *Eight days later* the Benelux countries declared themselves ready to enter into negotiations.

Strangely enough, our first contacts with the British Foreign Office were disappointing. Although Bevin's speech in Parliament was undoubtedly directed against the USSR, the British seemed most anxious not to irritate the Russian bear. When Hector McNeil, Bevin's Principal Assistant, came to see Spaak at the beginning of February, the two men were on quite different tracks. Those differences of opinion are reflected in three diplomatic notes exchanged between the Western Governments on 19 February, one from the Benelux countries, one French, and the other British.

The British and French documents indicated in similar terms that the two Governments acted in close concert. Both stated that nothing should be done that could upset or even worry the

Kremlin. They therefore proposed West European arrangements based on the Franco-British Treaty of Dunkirk, which would consist in a military alliance directed *against Germany*, and organised for a concerted action to prevent German aggression! The British and French documents also suggested *bilateral* agreements between the contracting powers rather than a multilateral pact. The idea, often expressed in British diplomacy at the time — an idea which we in Brussels never understood; was that the Americans were opposed to Western Europe getting together and that, if it did so, the Americans would feel that European defence was assured and Washington would therefore be encouraged to withdraw its forces.

Spaak, like most European statesmen, was of the opinion that no European defence was possible without American participation. But he was convinced that the more united the Europeans were, the more confidence they would inspire on the other side of the Atlantic. On the other hand, he felt strongly that a firm and definite attitude would restore to the peoples of the Continent the self-esteem and courage that years of suffering under enemy occupation had badly shaken. He also believed that it would be a unique deterrent against Soviet adventures.

Bech in Luxembourg and van Boetzelaer in The Hague endorsed Spaak's views. The Benelux memorandum of February accordingly rejected all diplomatic subtleties and proposed in bold terms (very much inspired by the Belgian note of 1944) a European organisation based on the United Nations Charter, which had provided for the possibility of pacts for regional security; it cited as an instance of what was wanted the Inter-American Treaty of mutual assistance signed in Rio de Janeiro in September 1947. Mutual assistance in case of aggression — any aggression — should be automatic and immediate: a system of regular and periodic consultations on all subjects of common interest was therefore required. In fact, the Benelux memorandum proposed a triple multilateral agreement: political, military and economic.

In spite of these initial differences, the Five powers agreed to enter into negotiations at once. A few weeks later, on 17 March (with extraordinary efficiency, achieved because of the fears inspired by the Prague coup and the Soviet threats against Norway), a treaty for a duration of fifty years was solemnly signed in Brussels between France, Britain, the Netherlands, Luxembourg and Belgium. While it still expressed concern over a possible German threat, it went far beyond that. It created for the first time in history a permanent political and military organisation in Western Europe against any foreign aggressor, obviously meaning the USSR. And it provided, also for the first time, a permanent Council for consultations at the highest level which could meet at any time for discussion of any problems — including economic ones — of common interest. Many essential requests in the Benelux memorandum were embodied in the new Pact.

One might ask: "Why so much about the Brussels Treaty in an article on the Atlantic Alliance?" The answer is that there would have been no Atlantic Alliance, I believe, if there had not first been a European Alliance. The immense difficulties faced by the American Administration have been explained by Ted Achilles and others in earlier chapters — how to transform a tradition of more than 100 years of isolationism into a quasi-permanent commitment on the continent of Europe. How to convince the American nation that it was in its interest to keep, for decades, hundreds of thousands of its young men some 5,000 miles away from home to preserve European freedom? It could be done only if the Europeans showed themselves ready to take the lead. The very day the Brussels Treaty was finalised, Spaak had a transatlantic telephone conversation with (if my memory is correct) President Truman himself; the President told him, "Now I am able to go ahead." The very day the Treaty was signed, the President pronounced before the United States Senate the celebrated speech which paved the way for the Atlantic Pact.

Other contributors to this book have described the many obstacles that still had to be overcome before the "miraculous birth of the North Atlantic Alliance" could become a reality. None of these obstacles was created by the Belgian Government. The Brussels Treaty had been approved by an overwhelming majority in the Belgian parliament. The Communists alone — and there were only a few of them — had voted against it, a remarkable achievement in a country which for more than a century had believed that neutrality would guarantee security.

Feelings were no different about an alliance with the United

States — rather the opposite. The man in the street, as well as the main political parties of both right and left, had become keenly aware that no defence was possible without America; that Britain, at the time the only military power in Europe, could not by itself resist the colossal armies of the USSR, which were still on a wartime footing. As the Berlin blockade had just begun, a complete commitment of the United States seemed of paramount importance — a deep concern completely shared by Paul-Henri Spaak and by most of the top officials of the Belgian Foreign Ministry. It was therefore with something near to amazement that I read in Ambassador Reid's account (see page 80, above) that, in the summer of 1948 Ottawa had formed the opinion that Spaak was "turning over in his mind whether an armed neutrality of West European countries might not be preferable to a North Atlantic Treaty". I have never heard anything to substantiate this. On the contrary, Spaak was quite worried about the persistent hesitations in Washington and the many delays in achieving an agreement. He used all his energy and talent to bring it about. It was in September of that year that he delivered his remarkable Paris speech which became for a while the clarion-call of resistance against the USSR. In the following month he strongly supported the motion of the Council of the Brussels Powers calling publicly for "a defensive Atlantic Pact".

When the NATO Treaty was eventually signed on 4 April 1949, Spaak considered it, not without reason, as something of a personal success. The day the Belgian Parliament approved it was one of the happy days of his professional life, as well it might be. For more than thirty-five years the Atlantic Alliance has secured "peace with honour", a rare achievement in the tormented history of Europe. In the small group of great men on both sides of the Ocean who made it possible, Paul-Henri Spaak is entitled to a respected place.

ABANDONING NEUTRALITY
How and why the Netherlands joined the Atlantic Alliance

Paul van Campen

During the greater part of the nineteenth and in the twentieth century up to the Second World War, Netherlands foreign policy was commonly known as one of neutrality or independence. This policy was due to a unilateral act; it was proclaimed and maintained by successive Netherlands Governments of their own free will and was neither instituted nor sanctioned by any international treaty or convention. It follows that the Dutch were entirely free to continue or not to continue this state of affairs — which explains why, particularly after the First World War, some people preferred to characterise the country's foreign policy as one of independence rather than neutrality. Now, the question arises: what was the basis of this policy?

Centuries of European history had repeatedly shown that the possession of the Netherlands, the delta of some of Europe's most important rivers, was a pre-condition of continental or world supremacy. Indeed, more generally the country's strategic position, its wide range of important colonial possessions and its commercial power went a long way to explain why the two great Continental powers could never allow the permanent domination of the Netherlands by either one of them; and Britain, in particular, always felt that the independence and inviolability of the Netherlands was a necessary condition of her own security. When, therefore, Dutch power and resources weakened in the eighteenth century, it was natural that Dutch diplomacy should turn increasingly towards a policy of non-involvment and political neutrality. Nethertheless this did not become a permanent feature of the international situation until after 1839.

The Dutch colonial possessions have already been mentioned. In fact, from early in the nineteenth century, the Netherlands was a relatively small power in Europe, but the case was very different in the Far East, where it possessed a colonial empire of the utmost importance. However, the country was unable, or unwilling, to defend this colonial empire against the dominating powers: so here too neutrality seemed natural, the more so since a similar state of conflicting ambitions as in Europe could be exploited; and for the time being it suited the world powers to leave the East Indies in Dutch hands. In addition to these geopolitical factors, it is also clear that as a mercantile state the Netherlands was likely to find a policy of neutrality attractive. In the course of time, this policy assumed all the respectability of a heritage handed down to successive generations by the wisdom of their ancestors; it had been tested and, it was thought, found good; and in due course tradition itself became one of the policy's most respected supports.

Inevitably the effect of such a long-standing policy is that the patterns of thought and conduct to which it gives rise continue to exist below the surface long after the formal disappearance of the policy which contributed to their rise. We have seen the reasons why the policy of neutrality became attractive; but highly-qualified observers saw other factors which worked in the same direction. Since the eighteenth century, the Dutch have been noted for their inclination to pacifism and their particular interest in the development of international law.

These tendencies most certainly did not grow out of the policy of neutrality itself: the name of Grotius and Frederick the Great's dictum — *"La Hollande est pacifique par nature et guerrière par accident"* — bear witness to this. But there is equally no doubt that neutrality strongly encouraged such patterns of thought and conduct. And this is of course quite natural: a small and relatively weak power has an interest in promoting causes which might be expected to ensure the security of states like itself. However, it would be vain to deny that the Dutch indulged a little too freely in these inclinations. They often overrated the importance and force of international law in the world arena; they sometimes quite wrongly regarded their own pacifism as part of a worldwide trend. Their preoccupations were no doubt admirable but they took them much too seriously. Even worse, these beliefs added to a policy of passivity and observation rather than activity and participation, led to a deplorable degree of apathy within the nation regarding problems of foreign policy and, worst of all, the military defence of the country. In short, when the hurricane of 1940 approached, the country had become dangerously detached from international realities and was unprepared mentally, morally and militarily.

After the tragedy of the war and the German occupation, there was widespread agreement that neutrality stood condemned. Was this because 1940 had clearly demonstrated that neutrality had failed as a guarantee of security? Very probably; but if so, the argument was based on an illusion — widely shared, but still an illusion. After all, even before 1940, it should have been clear that neutrality could never have offered real security to the country. This policy, so useful in times of an accepted balance of power, was bound to fail if and when that balance was threatened or destroyed.

In the event of one of the great powers initiating a policy of expansion, setting itself the definite goal of continental or world domination, the neutrality and independence of the Netherlands could not possibly be respected, precisely because of the very strategic and economic basis upon which, in times of equilibrium, a policy of neutrality could be based. Seen in this light, the events of May 1940 could not have been very surprising. Still, the fact is that many people were extremely surprised, pained and shocked, which can only be explained by the fact that the country escaped the miseries of the First World War. Yet while most Dutch people ascribed this to the merits of neutrality, the historical truth is quite different: that fortunate escape was mainly due to the pure accident that a strong and competent Chief of Staff in Germany (Schlieffen) was succeeded by a weak and incompetent one (Moltke).

In the event, there were solid arguments for thinking that, whatever the merits of neutrality in the past, the security of the country in future should be ensured by different means. For example, it was by now generally realised that the emergence of long-range airpower had caused a fundamental change in strategic conditions. Also, the evolution in general technological conditions and the narrow industrial base of the country made it impossible to organise adequate military defence in isolation. In other words, the technical conditions and requirements of modern defence posed these alternatives: either the country prepared its defence in co-operation with others *before* an attack — which ruled out neutrality — or no defence would be possible at all. This, after the dreadful calamities of the Second World War, the nation was unwilling to accept.

However, the general political outlook in the early 1940s may well have formed the decisive argument. Neutrality depends on either genuine general peace, or at least on a global or continental balance of power. Could this be expected in the conditions of the 1940s? There could be no doubt that Germany was beaten, but Germany had been beaten before; was this in itself sufficient evidence that it would never rise again? Fear of German aggression persisted in the Netherlands for a long time after the war, to be succeeded later by the realisation that Russia was by far the greater danger of the two. But, whether Russia or Germany, the prospect of an undisturbed equilibrium was dim. The developments of 1945–8 confirm that this was, indeed, the decisive issue which led to the Dutch discarding that neutrality.

Against this background, there can be no surprise that the Dutch sought their future security first in the United Nations system. After all, the United Nations embodied the principle of universal co-operation; and the subsequent collapse of this principle was only hesitantly and reluctantly accepted by the Netherlands Government. This option was quite in accord with the somewhat superficial idealism so characteristic of the Dutch and their reluctance to face the world as it is. Nevertheless, there was no overwhelming sympathy for the structure of the United Nations as laid down in the UN Charter. While an exhaustive discussion of the Dutch view of the United Nations is outside the scope of this article, two points nevertheless merit special mention.

First, the Netherlands Government was far from happy with the proposed structure of the United Nations, and made a number of suggestions and amendments all more or less reflecting a somewhat legalistic approach and a sincere desire to find constructive compromises for existing differences and contrasting views — both elements being almost constant characteristics of Dutch foreign policy. The Dutch were particularly unhappy with the standard established by the Charter, namely the maintenance of international peace and security. They felt (impact of the tradition!) that the elements of morality and justice in international law should be brought in. They were equally unhappy with the right of veto. However, the Charter in its final form, as agreed upon at the San Francisco Conference, clearly shows that the efforts of the Netherlands Government were not particularly successful. This did not, of course, prompt it to reject membership, if only for the simple reason that it was in no position to do so.

Secondly, could Netherlands membership of the United Nations be reconciled with any form of neutrality? The Government denied this in 1945; on the other hand, in 1948* the Foreign Minister stated that neutrality was only definitely placed beyond the law in 1948 with the establishment of the Brussels Pact — a fact of special importance if one considers that this Pact was the direct forerunner of the Treaty of Washington in 1949.

We should not forget the position of the Netherlands in relation to the implications of its membership of the League of Nations as war approached. When it became clear that the League had failed, the Netherlands joined certain other nations in stating in Copenhagen that application of Articles 10 and 16 of the League's Covenant could only be facultative (July 1936). In other words, it was for the Netherlands Government to decide whether the Covenant had been violated or not, and whether, if it had been violated, application of sanctions should follow.

In 1946, in the course of the annual debates in the States General (parliament) on foreign affairs at the end of the year, the Netherlands Government made a statement on the general principles and lines of their foreign policy. The following were among the points dealt with. First, neutrality was rejected because it could no longer guarantee the country's security. Hence, the principal of collective security was adopted, and thus support for the United Nations despite its many structural imperfections. Universal co-operation was proclaimed to be the country's primary objective in international relations. But since the United Nations was still in its infancy, participation in regional arrangements might well become desirable in the future. However, this was only in regard to cultural and economic matters, and *not* security.

---

*Baron van Boetzelaer, speaking at the conclusion of the Brussels Pact in 1948, said: *"Toutefois, la neutralité n'était ni après la première, ni après la seconde guerre mondiale, mise hors la loi."*

As for the future, this would depend entirely on what happened to the main and essential aspect of universal co-operation, namely the co-operation between the four big powers. If, for instance, the German problem could be dealt with in a satisfactory, united way, the Netherlands would then be the neighbour of a militarily-weakened Germany. In that case there was little reason to be afraid of in becoming a military bridgehead in Europe, for the country's security would be assured by the United Nations, by a four-power agreement on Germany and, finally, by the fact that Germany would be disarmed. Thus the German problem formed the crux of the European problem.

At that time, the Government certainly did not deny that things were not going particularly well with four-power co-operation; but in 1946 the Dutch were not prepared to assume that disagreement between the powers would become a permanent feature of the post-war world. Let us note also that in these circumstances, the Netherlands did not show much enthusiasm for either the idea of a West European bloc or a federal Europe. Some of the reasons for this attitude, at least from the traditional Dutch point of view, were not difficult to understand. After all, the Dutch felt it was not in their country's interest to take any step, or prompt any ideas, likely to exacerbate the emerging division between Eastern and Western Europe. Again, what security value could be attached to regional military arrangements if the United States did not take part? That precisely was the attraction of four-power co-operation and a four-power agreement: the United States would automatically be involved.

However, the events of 1947 were to show that agreement between the powers was a rapidly vanishing possibility. The Moscow Conference in the spring of that year, which was supposed to deal with the German problem, ended in complete failure; there was also General Marshall's Harvard speech on 5 June 1947 formulating a conception for the whole of Europe; but since Russia refused to participate, and since she forced her satellites to follow her example, it became quite clear that the breach between the great powers was almost complete.

The reaction of the Netherlands Government to this evolution was extremely prudent. Nevertheless, by November 1947 a slight but unmistakable change in the Dutch attitude became obvious. It was now said that the policy of universal co-operation and, in Europe, a four-power agreement under the United Nations could only be maintained as long as the international situation did not show *fundamental* changes. For instance "should, contrary to our hopes, the division of Germany into Western and Eastern parts assume a more or less definite character, the Government would feel compelled to reconsider its whole policy in Europe."*

However, having said that, the Dutch Government still refused to discuss alternative policies, and persisted in its negative attitude to all ideas of European federalism or a West European bloc. It also deprecated suggestions made in the Second Chamber of the States General to establish closer relations with France and Britain; in this context, the "open" character of the Dunkirk Treaty concluded in March 1947 between France and Britain had been stressed in Parliament. In any event, Dutch policy opposed the very principle of bilateral agreements. It should be noted that this reluctance and hesitancy over taking decisions was not always properly understood and was sometimes mistaken for an absence of any Dutch policy at all. The continuing references to a "One World" policy irritated a good many people. However, the year 1948 was to see significant developments, and what might be called a period of transition in Dutch foreign policy was coming to an end.

Indeed 1948, an important year as a whole, opened on 22 January with a significant speech by Ernest Bevin, the British Foreign Secretary who said *inter alia*:

"I believe that the time is ripe for the consolidation of Western Europe. First in this context we think of the people of France. The time has come to find ways and means of developing our relations with the Benelux countries. I mean to begin talks with those countries in close accord with our French allies. Yesterday, our representatives in Brussels, The Hague and Luxembourg were instructed to propose such talks in concert with their French colleagues. I hope *treaties* will be signed with our near neighbours, the Benelux countries, making with our treaty with France, an important nucleus in Western Europe.

"We have then to go beyond the circle of our immediate neighbours; we shall have to consider the question of associating other historic members of European civilisation, including the new Italy, in this great conception. . . ."

*Memorandum of Reply, Second Chamber, 13 November 1947, p. 14.

The Dutch reaction to this startling departure, with its unambiguous implications, illustrated that something had changed in The Hague as well. The reason was not far to seek. The failure of the London Conference in December 1947, which meant the definite breakdown of four-power co-operation, represented, in the terminology of November 1947, a change of fundamental importance. The time for "reconsideration" and a switch to Alliance policies had come.

The Brussels Pact is of quite exceptional importance, not only in its own right, but also because it was — and was already so seen at the time — to be the forerunner of a wider security pact, the Treaty of Washington of 1949. Within a week following Ernest Bevin's speech, the Benelux powers, at their conference on 29-31 January 1948 in Luxembourg, reached agreement on the principles of a common Benelux policy, thus initiating a form of political co-operation which was successfully maintained in all subsequent negotiations. The essential points of this common policy included the rejection of the principle of a network of bilateral agreements and a clear-cut preference for multilateral regional agreements based on Articles 51, 52, 53 and 54 of the UN Charter. Bilateral treaties, such as the Dunkirk Treaty, expressly referring to Germany, did not correspond to realities; a regional system, aiming at the consolidation of Western Europe, directed against no power in particular, was much more to the liking of Benelux, and certainly the Netherlands, and fitted more particularly into the traditions of Dutch policy. In the same order of ideas, the priority given to the economic and social aspects of the proposed system of co-operation should be noted: they were to be completed by arrangements for mutual military assistance in case of aggression.

Last, but not least, the Netherlands Government supported as a matter of course the British position as to the full utilisation of *overseas territories* in the planned Western European system of association — a point which was to create certain complications then as later. But this was not all. The Benelux Governments, if they were to assume obligations, felt that they should clearly have rights as well, particularly the right to be associated in the deliberations and decisions regarding a problem equally essential to all of them — Germany. Indeed, Benelux felt that without such association the proposed arrangements would be of little importance to them — a somewhat unfortunate exercise in bluff, and one which was rather quickly called when the results of the Six Power Conferences on Germany of February-June 1948 became known. These, for Benelux, were meagre.

Ultimately, in spite of the Franco-British predilection for bilateral treaties, it was Benelux's multilateral, regional principle which was accepted in the end as the basis for the Brussels Pact. As for the question of automatic military assistance, the Benelux countries had been reluctant to assume obligations of automatic assistance in case of aggression against one of the contracting powers *anywhere in the world*. The Dutch would also have preferred no country in particular to be singled out in the Treaty as a potential aggressor, but other powers had insisted upon this being done. Finally, the Dutch particularly welcomed the Consultative Council as one of the most important features of the Treaty. In their view, the Council should be seen as an "advisory body" to the Governments concerned. Via the Council, Benelux might hope to influence, to a certain extent, the policies of the great powers.

The treaty was signed in Brussels on 17 March 1948, and in his speech on that occasion, the Netherlands Foreign Minister, Baron van Boetzelaer, heavily underlined the special importance of the Pact for Netherlands foreign policy:

"The signature of the Western European regional Treaty is a milestone of great importance in the history of the Kingdom of the Netherlands. During the nineteenth century, my country maintained a 'policy of neutrality', which might more appropriately be termed a 'policy of independence' as the neutrality in question was not a permanent one guaranteed by the Great Powers. The Netherlands regarded neutrality as the guiding principle of its international policy, believing that it helped to promote a proper balance among the great European Powers which, it was convinced, was a decisive factor for peace in Europe.

"Theoretically, this policy of independence was abandoned when my country joined the League of Nations and, subsequently, the United Nations. However, neutrality was not repudiated in the wake of either the First or the Second World War. It was still quite possible to remain neutral should the Great Powers disagree as to who was the aggressor. The Western Union Pact we are signing today means that there can no longer be any question of remaining neutral if one of the five partners is the subject of aggression in Europe; it also guarantees that should we ourselves be attacked, we shall receive all necessary military and other assistance.

"Her Majesty's Government is convinced that the signature of this Treaty by the Netherlands will be supported by a very large majority of our people. . . . As Mr Bevin has said, it provides a 'nucleus' which other States, actuated by the same ideal, will be able to join once it has taken shape and grown in strength. . . We shall greet with joy the day when the group of countries participating in our regional union is expanded; but before this aim is pursued we believe that the first task must be to deepen this union. . . Nor have I concealed from you the fact that we appreciate the gravity of this decision which, in some respects, represents a turning-point in my country's traditional policy."

The reception of the Brussels Pact in the Netherlands was on the whole very positive. It was generally realised that the Pact meant a definite break with the past; as long as the United Nations could not guarantee its security, it was clearly the country's duty in these circumstances to protect itself from aggression by way of complementary arrangements. The great majority of the States General supported this view, but the applause was not general, and the opposite view was stated in somewhat extreme fashion in the First Chamber. It advocated a new concept of neutrality, which turned out to have certain attractions in later years. That concept was roundly based on the fear of atomic arms and the folly of total war — preferring a policy of playing safe to being destroyed.*

As for those who, while agreeing in principle with the Pact and the policy it represented, criticised certain points, the following may suffice. During the parliamentary debates, distrust of British policies was clearly noticeable in many quarters. These feelings can be explained by certain events in the common history of the two countries, but they had been severely aggravated by certain elements in British policy over the Indonesian question — a problem to which we shall return. Some members also felt that the Pact was very vague on the question of a possible aggressor. It was regretted that the Dunkirk Treaty was not

---

*First Chamber, Pact of Brussels, 23 June 1948, 667 ff.

abrogated by the Brussels Pact. The Dunkirk agreement, after all, was clearly directed against Germany, and the Brussels Pact echoed this in Article 7. The Pact's ultimate aim was the economic reconstruction and reconsolidation of Western Europe. But it was contended that such aims were not likely to be furthered by such an unnecessary reference to Germany — which, moreover, went counter to what might be called the "European spirit". Was not the reconstruction of Germany a vital interest to the Netherlands? Did the fact that the old Dutch "policy of independence" could not be continued imply that we had to follow France and Britain in every respect? Was it, therefore, not true that the very terms "Western European Union" or "Western European Treaty" were incorrect, first because of the factor of overseas territories, secondly because it was not said anywhere that only Western European powers could accede to the Pact, and thirdly because the idea that the aims of the Pact could be realised by Western European powers alone was a dangerous illusion? *A "European policy of independence" was quite as illusory as its Netherlands equivalent had proved to be.* Consequently, accession of other powers to the Pact should be encouraged; nor should the possibility of complementary arrangements between the five powers of Brussels and other nations — particularly in the military field — be overlooked.

Finally, for the sake of completeness, it should be added that in the framework of the parliamentary discussions the Netherlands Government had to sustain a heavy attack by what might be called the European Federalists. This theme being outside the scope of the present article, I will merely point out that already at that time the thesis was heard that Europe should not be a satellite of either the United States or the Soviet Union. It was contended that Europe would have to change its existing organisation. This thesis was rightly rejected — rightly, because a possible Third Force role for Europe depended, and still depends, first and foremost on its power. It is nevertheless interesting to note that, like the theme of Europe being unable to go it alone, the opposite idea of the Third Force, which would come to the fore later on, was already on the table in 1948.

The Netherlands had always considered the Brussels Pact as a beginning. But at the same time, it felt that undue haste in expanding the Pact should be avoided. There were two reasons for this attitude: first, that there should be time for consolidation; and secondly, that certain forms of enlargement might actually result in a weakening of the Pact. The Hague felt, in fact, that there were countries whose accession would only extend the obligations of the other parties to the Pact, without adding anything at all to their military potential.

However, the relations of the Pact powers to the United States and Canada were in a quite different category. Both Government and States General fully understood that the backing of the United States was necessary to make any European association a political and military reality. The Vandenberg Resolution of 11 June 1948 made it possible to translate into fact the intentions of President Truman to assist the Europeans, announced to Congress in his 17 March statement on the day the Brussels Pact was signed.

The negotiations leading to the conclusion of the Treaty of Washington opened on 6 July 1948. There is no need to describe these negotiations yet again, but the following points are of importance. That the Netherlands welcomed these developments will be clear. The certainty was obtained that the United States was prepared to accept full responsibility in a system of common defence and security, not only in the field of military supplies, but also in that of military action. Nor did The Hague object to the fact that an altogether new Treaty would be required. The Brussels Pact, after all, had served its purpose — it proved to be the beginning of a wider association.

After further consultations and negotiations, the agreed text of the Treaty was signed on 4 April 1949. The North Atlantic Treaty, like the Brussels Pact, was, as far as the Dutch were concerned, the end of an illusion: the hope that the United Nations could, by itself, ensure international peace. However, the real moment of innovation was the 1948 Treaty of Brussels and *not* the 1949 Treaty of Washington.

The Atlantic Treaty was, as expected, generally accepted. Certain points were of specific interest to the Netherlands or acquired special significance during NATO's subsequent evolution. As in the parliamentary discussions on the Brussels Treaty, the Indonesian question proved especially difficult. The Treaty's success was inconceivable without the existence of friendly relations between the signatory powers, but, it was asked, did such friendly relations really exist between the Netherlands on the one hand and Britain, the United States and Canada on the other? Once again, reference was made to existing embargoes on arms deliveries; at the time of the Brussels Treaty, the British embargo was already regarded as an anti-Netherlands act. Now the question was asked whether such embargoes did not show unfriendly intentions and an equally unfriendly tendency to interfere in what was, at least according to the official Netherlands viewpoint, a purely internal affair.

In addition, there was a fairly general doubt in Parliament as to the implications of Article 4 of the Treaty, precisely on account of Indonesia. Did this Article imply that the Netherlands was obliged to consult all parties to the Treaty with respect to the Indonesian question? And what would be the position if one or more signatories to the Treaty accused another of endangering peace and security in its own territory?

Finally, the embargo question had other aspects as well. If it was true that the world situation required the re-armament of Western Europe, the question might be asked whether the Netherlands could consider the British and American embargoes compatible with Article 3 of the Treaty. In view of several statements made at that time in the United States concerning Dutch policy in Indonesia, the question was therefore put as to what the Netherlands Government expected in connection with Article 3. Although such questions are now purely historical, it is nevertheless of interest that the then Foreign Minister, Dirk Stikker (subsequently Secretary-General of NATO) re-affirmed the Netherlands' view that the Indonesian affair should be considered as a purely internal question and that Article 4 could by no means licence the contracting powers to meddle in each others affairs. As for the embargo question and Article 3, Stikker pointed out that the Treaty could only be based on the principle of equal treatment of all participating powers and that he had protested against the embargoes; but he added that it was not Netherlands policy to make the ratification of the Treaty dependent upon the repeal of these restricting measures. However, if the powers took measures which prevented the recovery of the Netherlands (a clear allusion to Germany!), or made the maintenance of law and order in the overseas territories and their development impossible, the Netherlands should not be blamed for possible consequences. The hope was therefore expressed that such interfering and discriminating measures would soon belong to the past, along with the situations which had given rise to them.

Article 5 of the Treaty was criticised for being less stringent and far-reaching than the comparable provisions of the Brussels Treaty. In reply, and in rebuttal of this criticism, Stikker compared the Brussels Pact, the North Atlantic Treaty and the United Nations system to a system of three concentric circles. The inner circle was, of course, the Brussels Pact, with its completely automatic aid formula; then followed the North Atlantic Treaty with its admittedly weaker mutual obligations; and, finally, the United Nations system in which the aid formula was still less stringent. The resulting security guarantees were, it was stated, not unsatisfactory; at any rate, Article 5 of the Washington Treaty was the best formula that could be obtained in the circumstances.

In the context of a possible enlargement of the Treaty, the position of various countries was again discussed in the States General — including Spain, for example. In the opinion of the contracting powers, Spain could not, for the time being, become a party to the Treaty. It was, however, put to the Government that if Portugal was acceptable, there seemed very little reason for excluding Spain. But this argument was not accepted on the grounds that the Security Council had seen fit to take certain measures in the case of Spain but not in the case of Portugal.

The possible accession of Turkey and Greece was also foreseen; and — a typical Netherlands remark — some doubts were expressed as to whether Turkey and the Western European nations really shared a "common heritage and civilisation". This observation revealed the strength of religious and other less relevant factors in Dutch foreign policy discussions, as was also shown by the considerable time spent in Parliament on the preamble of the Brussels Pact because it made no mention of the name of God.

Finally, a lot of time was spent in discussing the problem of the extent and range of the Treaty and the Organisation. The Brussels Pact, it will be recalled, covered the social, cultural and economic aspects of the planned system of the association in a fairly thorough manner: this was not the case with the North

Atlantic Treaty. For the Netherlands this point was of some concern, because it was generally felt that if the military assistance of the United States to the European nations was not formally based on already existing broad economic assistance programmes, it was pointless. The point was therefore made in parliamentary discussions, and confirmed by the Government, that the North Atlantic Treaty was "a necessary complement to the broad economic co-ordination now proceeding under the European Recovery Programme [ERP]". However, the absence of a formal link between the Treaty and the ERP was only because the latter included countries which would not participate in the Treaty. The point may therefore be considered of a somewhat academic interest only: after all, without the communion of interests, as symbolised in ERP, there would have been no Atlantic Treaty.

Nevertheless, the Dutch concerns found expression in a very practical point: it was then, and remains, the firm conviction of successive Netherlands Governments that the Council, referred to in Article 9 of the Treaty, should co-ordinate the political and military policies of the participating powers. In the final analysis, what the Dutch wanted in 1949 was an Atlantic Treaty with the adoption of such economic and social policies as might be expected to make the concept of Atlantic community a powerful reality, worthy of human faith and sacrifice. In short, the North Atlantic Treaty Organisation, instead of being merely a military expedient, should be positive in character and effect.

The day the Treaty of Washington was signed marked the culminating point of developments which had begun in 1945, or perhaps even earlier, with a speech by Van Kleffens (Foreign Minister in the wartime Netherlands Government in London) on 28 December 1943. In that speech, the fundamental choice between isolation or association was put before the nation. Indeed, many of the fundamental security considerations which were to come to the fore in the next decades were already clearly foreseen and explored on that occasion.

There can be little doubt that it was the security factor which dominated all other considerations in the evolution of Netherlands foreign policy from neutrality, via universal and collective co-operation, to membership of a regional defensive organisation. Successive Netherlands Governments of that period realised that security for the Kingdom in Europe was their prime responsibility: "Our first task is to see to it that our own country is protected in the best manner possible."* The fate of the collective security system embodied by the United Nations was, not unnaturally for a country in the geographic situation of the Netherlands, particularly illustrated by the degree of co-operation between the four great powers on European problems.

The Hague judged this whole question in the final analysis by the evolution of the German problem. Therefore, to simplify the matter somewhat, Germany was for the Netherlands a barometer of the general security situation. There is no need to explain the absolutely fundamental importance of Germany for the Netherlands in military, political and, above all, economic terms. The fact remains, however, that when it proved impossible to reach agreement on Germany and the division of Europe became an established fact, for the Netherlands the die was cast: it was realised at that point that any four-power cooperation was excluded and, with it, all reasonable expectations of a successful United Nations collective security system.

The Indonesian question bedevilled relations between the Dutch and the Anglo-Saxon powers. However this issue, important as it was for the post-war generation, was not allowed to obscure the primary question of security for the Kingdom *in Europe*.

The reader will have noticed in this somewhat cursory survey of four years' evolution of Dutch foreign policy various elements which may be reminiscent of more recent problems in the Netherlands and other Alliance countries. In this particular context, it seems desirable to stress the general point that the mere signing of Treaties, instituting a fundamental change in a given policy, does not necessarily mean a parallel change in underlying mental attitudes; and this will be particularly true if both attitudes and policies have existed and been respected for more than a century. Given recent developments, is it possible that the Netherlands may now be returning to the old policy of neutrality and pacifism? It is the thesis of J.L. Heldring† that

---

*D. Stikker in the Second Chamber, 4 February 1949.
†J.L. Heldring, "Continuiteit in onze geschiedenis?", *NRC — Algemeen Handelsblad*, 4 December 1979.

this may be so, but it is only fair to add that this opinion is not generally accepted. However that may be, one can only hope that Government and nation, in reflecting on present-day problems, will be mindful of the fundamental issues of security which should always be the prime concern of any foreign policy.

OPTING FOR COMMITMENT

Luxembourg consolidates its independence within the North Atlantic Treaty Organisation

Nicolas Hommel

"By participating in the negotiation and settlement of the North Atlantic Treaty, the government of the Grand Duchy has followed the path to which it was committed by signing the Brussels Treaty, guided only by the wish to replace the illusory guarantees of our neutral status and the ever faltering guarantees of the United Nations Organisation with the concrete guarantees of a system of mutual assistance and collective defence against all aggression."

In these words Joseph Bech, Minister for Foreign Affairs, presented to the Legislative Assembly in April 1949 the bill for the ratification of the North Atlantic Treaty. These words summarise, with startling brevity, the fundamental evolution of the foreign policy of Luxembourg since the start of the Second World War and at the same time bear witness to its wish to adapt this policy to the new facts and realities of the post-war period.

How did Luxembourg arrive at this position? To have a clearer grasp of this evolution, it is necessary to go back to 1867 and retrace the principal episodes of a perpetual and unarmed neutrality, marked nevertheless by a fragility which in retrospect suggests that it was not at any time able effectively to guarantee the country's existence and independence.

Perpetual and unarmed neutrality was imposed on the Grand Duchy by the Treaty of London of 1867 as a political expedient to put an end to an alarming test of strength between the France of Napoleon III and the Germany of Bismarck. As a result, the powers signing the treaty placed this neutral status under the protection of their collective guarantee. Although born out of the circumstances of the moment and created with a specific aim, the Constitution of the Grand Duchy institutionalised the principle in its first Article in 1868.

This collective guarantee was soon seen to be very tenuous. From the beginning, in the eyes of the British government, the collective guarantee only in fact entailed a "moral commitment" for the guarantor-states, and would only require the armed defence of Luxembourg if *all* the guarantors unanimously agreed upon it. However, the British government's interpretation, often refuted by legal experts and probably contrary to the interpretation of most of the other signatories, was never retracted and was even confirmed by the course of events.

The fragility of the system of neutrality, which was based only upon the law, was first flagrantly demonstrated during the 1914-18 war. After violating its guarantee, Germany could no longer be considered, politically if not legally, as one of the guarantors of neutral status. Moreover in 1917, with the advent of the Bolshevik régime, Russia repudiated outright all its commitments arising out of international treaties concluded by the Tsarist régime. Here was one more defaulting guarantor.

The Treaty of Versailles of 1919 was shortly to add to the erosion of conventional neutrality. Article 10 required Germany "to confirm the abrogation of the neutral status of the Grand Duchy". As this provision was *res inter alios acta** for Luxembourg, obviously it could not mean the abrogation of its neutral status. What must however be borne in mind is the clear and precise declaration of the wish of all the guarantor-powers of the Treaty of London — except Russia — to abrogate Luxembourg's neutrality.

In such circumstances, Luxembourg had to look elsewhere for a sounder basis for its security than that offered by a treaty which was not formally abrogated but had been considerably weakened. In pursuit of this aim, it applied for admission to the League of Nations, while arguing that it did not wish to abandon its neutrality, which was deeply embedded in its spirit. This application posed a serious problem. In fact, it meant reconciling the military commitments of the Covenant of the League of Nations with the conventional and institutional status of permanent and unarmed neutrality. With the help of the new concept of "differential neutrality", following a misunderstanding between the League of Nations and the Luxembourg

---

*"*Res inter alios acta alteri non nocere debet*": legal term meaning that a transaction between others ought not to prejudice another party — in this case Luxembourg.

government over the latter's willingness to respect certain obligations (economic and military sanctions, right of passage of troops), Luxembourg was admitted but without the benefits of special treatment. Admission was accompanied by the recommendation that its Constitution be brought into line with the obligations of membership, namely that Article 1, declaring that the Grand Duchy was a perpetually neutral country, should be deleted. Did this request not mean that the member-states of the League of Nations, and among them the principal signatories of the Treaty of London, considered the Treaty to be abrogated?

With the benefit of hindsight, the Luxembourg Council of State could legitimately have doubted that neutrality could be maintained in its original form after the First World War and after admission to the League of Nations, and could have considered that its *contractual neutrality*, which began in 1867, had over the years turned into a *voluntary neutrality* which had become a line of conduct in foreign policy. However, this *voluntary* neutrality could not relieve the Luxembourg Government of the need to consolidate the country's international position. Joseph Bech threw all his energy into the pursuit of this policy. He applied to be a party to the Locarno agreements, but was turned down. Not discouraged, from 1927 onwards he signed more than a dozen arbitration treaties. But with insecurity building up again in the face of the increasing inefficacy of the League of Nations, Bech endeavoured in the years leading up to the Second World War to obtain confirmation, from the former guarantor-powers of the 1867 Treaty, of the contractual neutral status of the country. His efforts were in vain. These powers avoided any renewal of their commitments.

Faced with this failure and the increasing danger of war, Bech tried in 1938 to obtain a simultaneous declaration from France and Germany guaranteeing respect for Luxembourg's neutrality. Despite persistent efforts, this initiative failed in its turn. So the Grand Duchy found itself on the eve of the Second World War without real or reliable guarantees, right in the middle of the probable battlefield. It was this situation which in 1939 led Prince Felix to say to the British chargé d'affaires that Luxembourg was "in a desperate position".

The invasion of the country in 1940 put an end both to contractual neutrality — insofar as it still existed — and to voluntary neutrality. From that moment the Grand Duchy abandoned neutrality in favour of belligerence. Recognised as one of the Allies, it associated itself with numerous acts uniting the Allies in their war effort. And besides these diplomatic initiatives, which were an unambiguous expression of the commitment of the Grand Duchy to the Allied side, more concrete steps in the military field only strengthened this commitment. This was how, in particular, the exiled Government decreed limited troop recruitment in case of necessity. This opened the door for a military contribution to the Allied side.

The international situation of Luxembourg between the wars had been marked by wavering and uncertainty about the continuance of neutral status and by the crippling juggling of a policy of neutrality with public opinion. At the end of the Second World War the tendency was towards a clearcut status and to giving the independence and territorial integrity of the Grand Duchy a new and effective foundation. When it joined the League of Nations, the Government had tried to maintain its neutrality while assuring a guarantee of collective defence for the country. This time its choice was clear. Hence the United Nations Charter was approved without reservations of neutrality. This step was ratified in 1945, *unanimously*, by the Luxembourg Consultative Assembly, since there was no normally elected parliament.

Neutral status had, over the years, given rise to an easygoing and egotistical state of mind. No effort, no sacrifice, either financial or physical, had been necessary for the defence of independence. Observance of international law had been supposed to ensure it. The new policy, however, was to appeal for international solidarity to which each nation, large or small, would contribute. Luxembourg public opinion was going to have to learn the meaning of this solidarity by experience.

Unanimous agreement had been reached over the abandonment of neutrality implicit in joining the United Nations. The same could be said for the introduction of compulsory military service which the government presented as arising out of the obligations envisaged in the San Francisco Charter. Whereas the Communists were hostile to any armed force other than a "people's army", the Socialist Party, now in opposition, questioned the obligations assumed and what they implied, and the "cost-effectiveness" of a compulsory armed force. As for the majority parties, the Christian-Socialist Party accepted military service somewhat half-heartedly, while the Democratic and Patriotic Party took the most positive stance with regard to the obligations arising out of the Charter.

Thereafter, international tension became alarmingly acute. The expansionist and subversive policy of the USSR roused public opinion and made people more inclined to accept the obligations of solidarity resulting from the new foreign policy pursued by Bech. Despite the lukewarm solidarity displayed during the military service debates, Bech was able, in this atmosphere of fear, to sign the reply of the Benelux countries to Ernest Bevin's speech of 22 January 1948, in which the three foreign ministers declared that they were ready to begin negotiations after defining "the outlines of an attitude inspired by the *feeling of solidarity* in Western Europe". The Brussels Treaty of 17 March 1948 went beyond the traditional model of an alliance in bringing about an unusual degree of solidarity by providing for obligatory and automatic mutual assistance in the event of aggression. Bech, intending to base the country's independence on a solid alliance, accepted this obligation of assistance, which touched a sensitive nerve of public opinion with regard to solidarity.

The Brussels Treaty, however, more than any previous Treaty, was incompatible with a status of neutrality, which therefore had to be abandoned before the Treaty could be ratified by Parliament.

This ratification gave rise to a major debate during which the Socialists had recourse to a rearguard action in favour of neutrality. After this, on the basis of a statement from Bech that the previous guarantors of neutrality, namely the British and French Governments, had declared themselves to be no longer bound by their 1867 commitments, they finally admitted "that the neutral status of Luxembourg has at present no support at all . . . and has broken down." As for the Communists, while admitting that the "policy of neutrality has not saved us from invasions" and that it "no longer constitutes a guarantee against future dangers", concluded quite contradictorily that it "would nevertheless remain the best policy for maintaining our independence". The vote for the abandonment of neutrality was thus passed unanimously, except for the Communist votes.

The parliamentary debates on the Brussels Treaty had thus disposed of the problem of neutrality. Their dominant theme was to be the implications of the undertaking given regarding military aid and assistance. Bech declared that he was unable "to give details concerning the means and extent of our possible contribution", and after affirming that this contribution would be established "only by taking account of what is reasonably possible for a small country like ours", insisted on proclaiming that "the people of Luxembourg will wholeheartedly contribute their share to the consolidation of the security of the member-countries." Once again he emphasised that "where Providence has put us, we cannot stay out of any conflict and so cannot remain neutral." He finally offered Parliament the "choice between fatal neutrality and active cooperation with a policy of European solidarity", stressing "that outside the Treaty, which is the starting-point for an organisation of the Europe of the future, lies isolation." After this appeal, Parliament ratified the Treaty unanimously, with the exception of five Communist votes. Hardly had this vote been passed when the prospect of by-elections in June 1948 brought up again the problem of assistance obligations, in other words the obligation to introduce compulsory military service. Differing views within the Government on the implications of the Brussels Treaty encouraged the Socialist and Communist Parties in their opposition to conscription. But Communist motions demanding a return to voluntary service were finally defeated.

It was in this climate — among arguments about economics, demography and military interests, in which a spirit little used to international solidarity could be discerned — that the Grand Duchy took part in the negotiations for the North Atlantic Treaty. This reserve, which the first symptoms of the cold war should have overcome, did not shake Bech's resolve to follow the direction of the country's international policy and to remain faithful to his generous conception of his obligations to the other members. Thus Luxembourg caused no difficulties during the negotiations. Having no specific interests to defend, the Luxembourg negotiator, Hugues Le Gallais, Minister Plenipotentiary in Washington, was able to keep a low profile.

At the exploratory talks during the summer of 1948 in

Washington, Luxembourg was represented by the Belgian Ambassador, Baron Silvercruys. Luxembourg's representative began participating only at the meeting of 10 December 1948 which marked the start of negotiations. For the purposes of these negotiations, the five countries of the Brussels Treaty had drawn up a memorandum containing a passage of great importance from Luxembourg's point of view. This stated "that it was particularly desirable that the agreement should provide for measures of material assistance as quickly as possible in the case of armed attack, including *individual military assistance* on the part of *each* member accepting the full obligations of the Treaty. . . ." As this commitment to assistance inspired by the Brussels Treaty was considered too stringent, however, the United States and Canada endeavoured to reduce its severity by proposing that an element of judgement be introduced into Article 5 and that the reference to military action be eliminated. It was at this point in the negotiations that the Luxembourg spokesman maintained that such a non-binding reading of the key article of the Treaty ran the risk of having no deterrent effect on the Kremlin. The formula finally adopted was such as to reconcile the requirements for security, the duty of solidarity and the modest means of the country.

On other important controversial issues such as the admission of Norway and Italy, and the demarcation of the zone covered by the Treaty or the duration of the Treaty, and faced with the differing interests of neighbouring and friendly countries or American and Canadian reticence, depending on the problem, Luxembourg was very reserved. Overall, it could be said that the Luxembourg delegate expressed a favourable attitude to Italy early on; did not raise the slightest objection to the admission of Norway; shared the wish of France to include the French *département* of Algeria in the strategic zone; and argued for a treaty of a duration that "should be such as to contribute to confusion and uncertainty in the minds of the leaders of the Kremlin" faced with the resolve of the powers signing the Treaty to defend themselves.

When submitting the North Atlantic Treaty to Parliament, Bech, after noting the powerlessness of the United Nations to ensure world peace, emphasised the two basic elements of the new foreign policy, namely the emergence from an "isolation which was never splendid" and "replacing the illusory guarantees of our former neutral status with the concrete guarantees of a common and reciprocal defence system". With the Communists in mind, he laid the stress on the peaceful aspirations of both the Brussels Treaty and the North Atlantic Treaty, the latter being only the "corollary" of the Brussels Treaty and the "logical complement" to the Marshall Plan.

The Council of State's verdict was very favourable: "The North Atlantic Treaty gives our country concrete and reliable guarantees to safeguard its freedom and to maintain peace in general. Such advantages are well worth a few sacrifices." For the centre parties of Parliament, the Treaty would "give Luxembourg additional security, and contains full provisions to safeguard her independence and the freedom of her subjects." The parliamentary debates, which took place in an atmosphere of great calm, offered the Communist spokesmen one last chance to let off steam. The Government was odiously labelled a "warmonger, a hypocrite, a slave of American capitalism". Apart from this diatribe, criticism was aimed particularly at Article 5 which "bound the small to the large"; at the irresponsibility of the Government which was sacrificing youth to American strategic aims; and at the inconsistency of the Treaty with the United Nations Charter. One spokesman, after justifying the bilateral pacts concluded by the USSR with her neighbours, denounced the Atlantic Treaty as a "war pact" involving the country in "adventurist politics", while "a reasonable policy" had been abandoned with the abrogation of neutral status which "despite a few mishaps had stood the test of time".

The Socialists ratified the government's foreign policy, although not with "one hundred per cent enthusiasm". The virulent opposition of the Socialist Party and the press to military service had calmed down. All that remained was a simple request for information on the exact situation of the army within the framework of the new foreign policy. However, according to the Socialist spokesmen, the Treaty should have been open only to democratic countries and should not be an end in itself; the aim should be the reorganisation of the United Nations, the solution of the German problem, and general disarmament. Bech, concluding the debate, confined himself to refuting the

Communists' arguments. His answer to the question of the contribution to the Treaty was again partly evasive. According to him, the Grand Duchy's contribution would probably be of an economic nature. This lack of precision subsequently led to a continual questioning of the extent and nature of the country's contribution to the Alliance. The Treaty, like the Brussels Treaty, was finally adopted unanimously, except for the Communist votes.

Armed with this ratification, Bech was legitimately able to consider that he had consolidated the country's independence, and that he had taken her out of isolation, thrown off the straitjacket of neutrality and initiated a foreign policy which was perfectly integrated with the development of international relations. The political and economic foundation on which the policy of European integration was to be based was thenceforward assured.

AN UNARMED NATION
Iceland rides out the storm

Olafur Egilsson

Few events, if any, in the recent history of Iceland have caused such intense and bitter political conflict as the decision in late March 1949 to participate in the foundation of NATO. Only some four years previously — on 17 June 1944 — the Icelanders had restored their Republic. Earlier in the century, in common with many other nations, they had hoped that a declaration of neutrality would suffice to keep them outside Great Power conflicts. By the end of the Second World War, however, attitudes had changed in the light of all that had happened. These were uncertain times and it was clear that a small unarmed nation, which wished above all to avoid becoming subordinate to others and losing its newly-won independence, had difficult decisions to make. It was both the strategic importance of the country due to its geographical position and the dangers which appeared to threaten the independence of small nations which chiefly made it possible and necessary for Iceland to consider joining a military alliance. The fact that such an alliance appeared at first sight hardly the place for an unarmed nation seemed less important.

Iceland lay outside the zone of hostilities during the First World War, and yet the war was hardly over before it became apparent that this situation would probably change. At the 1920 World Congress of the Communist International, Lenin spoke of "the strategic position of Iceland in a future war, with particular reference to aerial and submarine warfare". The German geopolitician Karl Haushofer later likened Iceland to a pistol continually aimed at Britain, the United States and Canada. By the 1930s, advances in aviation had been such that Iceland was being examined more and more closely as a stepping-stone on air journeys between the continents of Europe and America. Attention was drawn to this development by, for example, the Italian Aviation Minister Balbo's flight of 24 military seaplanes across the Atlantic via Iceland in 1933.

Indeed, interest was already being shown in the New World in securing landing rights in Iceland for Atlantic postal flights.

Early in 1939, with the Second World War imminent, the German airline Lufthansa made a determined but unsuccessful attempt to obtain a licence to fly to Iceland. German submarines visited the port of Reykjavik and the widening sphere of submarine operations directed increasing interest towards Iceland. The importance of meteorological observations in Iceland for weather forecasts in the nearest countries and on the continent of Europe, as well as in the sea areas, became increasingly evident. The Icelanders had admittedly suffered shortages resulting from supply difficulties during earlier conflicts, for instance during the Napoleonic wars, but they had not become directly involved, remaining unarmed in the far distance. This had happened in the First World War, but they now had to face the fact that they were within the zone of the Second World War. The blessing of improved communications with the rest of the world was soon mixed by the curse of armed conflict which was later to cost the lives of many Icelandic seafarers.

When the Icelanders gained independence from Denmark in 1918, they made a solemn declaration of perpetual neutrality. But the hope that this neutrality would continue to protect them faded very quickly when war broke out again. During the first spring of the war, it was feared that Germany would occupy the country, but Britain beat them to it. Winston Churchill, then First Lord of the Admiralty, issued the following instructions at the end of April 1940, a few days before he became Prime Minister: "In view of the bad reports from the Faroes about aircraft or seaplane bases and the fact we must reckon with the Germans all along the Norwegian coast, it seems indispensable that we have a base in Iceland for our flying-boats and for oiling the ships on the Northern Patrol. Let a case be prepared for submission to the Foreign Office. The sooner we let the Icelanders know that this is what we require the better."

British forces landed at Reykjavik on 10 May 1940, thereby breaching Iceland's neutrality, and a formal objection was lodged. When Britain had made preliminary soundings a month earlier, the Icelanders had replied that they neither would nor could take part in military operations or enter into an alliance with any party to the hostilities. Nevertheless, once the occupying forces had arrived, both the Government and the people reckoned that what mattered most was that they had to deal with the combatant they would have opted for, and thus they gave the British troops a generally warm reception.

The United States assumed the defence of Iceland a year later in early July 1941, under the terms of an agreement with the Icelandic Government which the British had brought about. Britain had strongly urged the Icelandic Government to solicit American military protection, since the US felt that it could not send troops to Iceland without having first received a formal request. By taking over the defence of Iceland, the US lightened the burden on Britain.

The fact that the Icelandic Government agreed to ask for protection was taken in some quarters as indicating the demise of the policy of neutrality, but this was emphatically denied in the debate on the agreement in the Althing. It was pointed out, for example, that entrusting national defence to a neutral state — as the US then was — instead of having combatant troops in the country, represented less of an interference with Iceland's neutrality. Many people, on the other hand, were beginning to realise that the days of neutrality were numbered, whether they liked it or not.

This 1941 defence agreement with the United States was, in fact, Iceland's first major foreign policy decision after she assumed full responsibility for her own foreign affairs in April 1940. When, in 1918, she had gained independence, while remaining in monarchial union with Denmark, it had been agreed that Denmark would look after Iceland's foreign affairs by special authority, but it was no longer possible for the Danes to discharge this duty after the occupation of Denmark on 9 April 1940.

The decision to open negotiations for the stationing of foreign military forces in their country was not taken lightly by the Icelanders. The fact that such a short time had passed since they had regained their freedom made the preservation of that freedom all the more important. But in these uncertain times, Iceland's leaders thought it right to sacrifice the lesser interests for the greater — to put up with the discomfort of having foreign forces in the country in order to ensure that the nation's independence would be honoured and protected. Communists, concerned about the fighting on the newly-opened Eastern front (Germany invaded the Soviet Union on 22 June 1941), saw a ray of light in the agreement, in that this could make it easier for the Allies to come to the aid of the Soviets. They proposed that the Soviet Union, the United States and Britain be requested jointly to guarantee Iceland's freedom and security. Thus they tried belatedly to drag the Soviets into the debate, but in this they were unsuccessful. All the members of the Althing with the exception of the Communists later voted for the agreement, i.e. 39 in favour with three against. Conditions governing the stationing of American troops in Iceland were to include: an understanding that they would leave as soon as the war was over; recognition of Iceland's absolute freedom and sovereignty; and an undertaking not to interfere in the government of Iceland either then or later. This the United States accepted. The agreement on military protection and the stationing of troops in Iceland was not revised when the United States formally entered the war, and the situation thus remained unchanged in this respect until after the war had ended.

Iceland proved of great value to the Allies throughout the war, and the number of American troops based there reached some 45,000. The bases in Iceland were vital for the protection of men and supplies, including lend-lease matériel for the Soviet Union, being conveyed across the Atlantic. Iceland "guarded a lifeline over the Atlantic", as Churchill was later to put it at the Yalta Conference. The theories developed earlier in the century about the country's strategic importance had been vindicated.

The unequivocal desire of the Icelandic people at this juncture to stay detached as far as possible from squabbles and hostilities between nations was again clearly demonstrated when the Republic of Iceland, which had been re-established as soon as the Act of Union with Denmark made this possible on 17 June 1944, rejected the offer of the three Great Powers of the Yalta Conference of becoming a founder-member of the United Nations Organisation if she declared war on the Axis powers.* The spirit of neutrality was still abroad and the Icelanders had a clear idea of what was best for a small, unarmed nation, although they were nonetheless perfectly aware of what ideals they wished to uphold.

---

*Iceland joined the United Nations on 19 November 1946.

After the end of hostilities in the autumn of 1945, the US Government put forward proposals for a long-term lease (understood to mean for 99 years) for military bases at three locations in south-west Iceland. Many Icelanders found this request unreasonable, and it became the root cause of long-standing mistrust towards the United States. The idea of agreeing to the establishment of permanent military installations by a foreign power — albeit a friendly one — was completely alien to a newly independent small nation now that a state of peace seemed to be in the offing. All Iceland's political parties agreed to turn down the request and it was withdrawn. The last American soldiers left the country in April 1947.

An agreement had, however, been concluded in the autumn of 1946 allowing the US Government limited use of the airport at Keflavik to service the army of occupation in Germany, at the same time as the Icelandic Government formally took over the ownership of, and jurisdiction over, the installations of Keflavik which had been built during the war. An American contractor assumed responsibility for the operation of the airport over the next few years, and facilities for the general public were improved. The agreement was to last for the duration of the United States' obligations in Germany, but could be terminated unilaterally after a period of six-and-a-half years from its signature. This agreement was the subject of vigorous debate in Iceland. The coalition government comprising the Independence Party (conservative-liberal), the Social Democrats and the Socialist Unity Party (Communists), which had been formed to foster economic reconstruction after the war, split over the question of the airport, but the agreement was nonetheless passed by a 32–19 majority of the Althing.

Ominous developments which cast an ever-lengthening shadow over the post-war peace were closely followed by the Icelanders. In common with other free nations, they felt increasingly uneasy about the security of their country in the light of the Stalin Government's broken promises and of Communism's expansionist policies.

The Government of Iceland in the period 1947–9 was a coalition of three political parties, the Social Democrats, the Progressives (agrarian-centrists) and the Independence Party, with the fourth party, the Socialist Unity Party, in opposition. Some members of the Government discussed the idea of a defence alliance of Western nations in connection with a meeting of Nordic premiers in February 1948, to be attended by Prime Minister Stefan Jóhann Stefánsson, leader of the Social Democrats. The Icelandic ministers' first reactions to this idea were naturally cautious, such questions having been the most delicate and demanding in recent Icelandic politics. They nevertheless agreed that "there could be no objection if [the Prime Minister] were to canvass Nordic opinion on this question, but it would clearly not be possible at this stage to state what Iceland's attitude might be, should she be asked to do so." Iceland has particularly close bonds of affinity with the Scandinavian nations, and it was thus natural to explore their views first. The feeling that careful consideration should be given to the defence of the Nordic countries proved to be general at the Nordic summit, and the Ministers agreed to discuss the subject further. The gravity and importance of these questions became still greater when the Communists seized power in Czechoslovakia in February 1948. The alarm which this generated had its effect on the Icelanders too, as they realised more and more clearly that their security depended on the course of events on the continent of Europe. Nor did the blind belief of some of those in Iceland who shared the belief of the Kremlin leaders in the virtues of Communism make for greater peace of mind or optimism.

When Ernest Bevin explained his concept of a Western defence alliance in March 1948, he already numbered Iceland among those states which should become members. It later emerged that the Norwegians also considered this most important as did the United States, particularly in order to safeguard communications via the North Atlantic sea-lanes. On the other hand, Paul-Henri Spaak of Belgium said in November 1948 that the speedy establishment of the Alliance was more likely if the Alliance were confined to the seven states which had been involved in the preparatory talks; attempting to persuade Ireland and the Scandinavian countries to participate would only cause delay. He was, however, soon to defer to the Americans in this respect.

It was at about this time — the autumn of 1948 — that it first became generally known that Iceland might be given the opportunity of becoming a member of this Western Alliance in the making. Frantic articles by opponents of Western defence co-operation, particularly the Socialist Unity Party, soon started to appear in Icelandic newspapers. Some intellectuals also showed fierce resistance, mainly on grounds of nationalism. The next few months witnessed a violent propaganda campaign opposing the very idea of a Western defence alliance and accusing the Icelandic Government of shackling the nation in martial chains and of stripping her of her newly-won independence.

The American Ambassador in Reykjavik, Richard P. Butrick, advised the Minister for Foreign Affairs, Bjarni Benediktsson, in the strictest confidence on 7 December 1948 of the intention to include Iceland among the first states whose participation in the North Atlantic Alliance would be invited as soon as consultations between those parties already involved had reached the stage when it was considered advisable to approach others. The Government had every reason to treat the matter with caution, and requested more detailed information before a formal invitation was issued and published. The Foreign Minister pointed out in a discussion with the American Ambassador on 11 December 1948 that American military bases in Iceland were out of the question. That had been settled once and for all in the formulation of the 1946 Keflavik airport agreement. A confidential Note confirming that Iceland would be given the opportunity of joining the Alliance was subsequently delivered on 5 January.

The Icelanders laid emphasis on examining all aspects of the proposal, in particular the attitude of the other Nordic nations; on explaining Iceland's special circumstances; and on finding out more about the conditions attaching to membership. The Prime Minister, Stefán Jóhann Stefánsson, made use of his close links with his colleagues and fellow Social Democrats in Denmark, Norway and Sweden, namely Hans Hedtoft, Einar Gerhardsen and Tage Erlander, in order to keep abreast of developments. But the greatest burden rested on the broad shoulders of Bjarni Benediktsson, Iceland's forty-year-old Foreign Minister, a former professor of law and Mayor of Reykjavik, who had taken up his first ministerial appointment for the Independence Party only a year earlier. He undertook a visit to Scandinavia for detailed discussions with his colleagues and other influential figures in late January 1949.

The fact of her geographical positon far from the other Nordic countries was one of the reasons why Iceland never participated directly in the attempts by Denmark, Norway and Sweden in late 1948 to establish a Nordic defence alliance. But in view of this possibility of forming an association of closely-related states, it was both natural and essential for Iceland to find out whether an acceptable conclusion could be reached. The Norwegian suggestion that Iceland would take part in the final stages of the talks was, however, rejected by the Swedes, who reckoned that this would lead to too much discussion about the Atlantic Alliance. The talks between the three Scandinavian countries broke down at the end of January 1949, mainly because of Norwegian insistence on a prior guarantee that the United States would supply them with arms and undertake to come to their aid in case of an attack. Such links with the United States were inconsistent with Swedish foreign policy. It thus became clear that Denmark and Norway would take part in the foundation of NATO. The fact that Iceland would probably be pursuing the same course as these other two Nordic states made it much easier for her to make up her mind, particularly as such a narrow-based Nordic union would never have been strong enough to guarantee Iceland's defence and would thus have been a totally unrealistic solution to her security problems.

The Soviet Government cautioned the Norwegians about joining the Western Alliance, and in a private conversation with Bjarni Benediktsson at the end of January, the Norwegian Foreign Minister, Halvard Lange, spoke of his fear that their powerful neighbour in the East, which shared a common boundary with Norway, would move against them before the Western nations had had a chance to complete the formation of their Alliance.

The American Secretary of State, Dean Acheson, stated in a telegram to the Icelandic Government of 29 January 1949 that the United States had no desire to station troops in Iceland except in the gravest emergency. This knowledge came as a great relief to the leaders studying the subject, but as they found the whole affair still needing some clarification, little was said publicly about their deliberations at this stage. On the other hand, the debate continued with bitter argument both in the press and at public meetings. Opponents of the proposed membership claimed, for example, that Iceland's independence would count for nothing in such an association of large and powerful nations — that the country was being put up for sale. Iceland was becoming enmeshed in a military machine actively planning to attack the Soviet Union, and would be dragged into armed conflict whether she liked it or not. Some still spoke of neutrality. But it gradually became clear that the great majority of Government supporters favoured participation in Western defence co-operation, provided that sufficient note were taken of conditions peculiar to Iceland.

The question had now reached the stage where a decison would soon have to be taken, but it was still essential to have all the arguments clearly spelled out. The Foreign Minister thus proposed a fact-finding mission comprising three members of the Government — one from each party — to Washington. This was agreed, and the mission became a key factor in the Icelandic Government's final deliberations on the matter. The delegation, consisting of Bjarni Benediktsson for the Independence Party, Emil Jónsson for the Social Democratic Party and Eysteinn Jónsson for the Progressive Party, held talks with representatives of the United States Government in the period 14–17 March. All aspects of the subject were most carefully examined. In their report on the visit after their return, the Ministers disclosed that the US Government had made a declaration confirming:

(1) that in the event of war breaking out, the Alliance nations would request facilities in Iceland similar to those which had been made available in the last war, and that it would be entirely Iceland's own decision when these facilities would be made available,

(2) that all other parties to the agreement would fully understand Iceland's unique position,

(3) that it would be recognised that Iceland had no armed forces and had no intention of establishing any, and

(4) that foreign troops or bases in Iceland could not be considered in peace-time.

The question of whether it would be possible to incorporate the reservations which Iceland wished to make on account of her special circumstances into the text of the Treaty was raised during the discussions, but it was ruled out on the grounds that it could lead to other nations also wishing to incorporate reservations.

A formal invitation to Iceland to become a founder-member of the Alliance was handed over at the end of the talks, together with the text of the Treaty, and because it had been decided that the Treaty should be signed in early April 1949, there was no time to be lost. It even looked as if the time available to bring the matter to a conclusion might be too short. A decision, therefore, had to be taken. In the light of the facts now at their disposal, the Government decided that there were adequate grounds for Iceland to join this defence Alliance, of which the aim was to safeguard peace, freedom and democracy. Its basis was Article 51 of the Charter of the United Nations, the UN itself having failed to gather the necessary strength to ensure world peace. The Government laid a draft Resolution before the Althing on 28 March 1949 proposing membership of the new Alliance.

The closing stages, especially the final three-day battle, were particularly gruelling and exacting. When the Government's policy was made public, the Communists proposed a motion of No Confidence which was argued in a furious broadcast debate on the same day that the formal Resolution on membership was tabled in the Althing. The debate on the Resolution itself began on 29 March and continued the next day. The ensuing events were unparalleled in the history of the Althing. The Alliance's opponents, with the more fanatical Communists in the van, called a mass meeting near the House of the Althing in the early afternoon of 30 March. With emotions at fever pitch, it was now feared that an attempt would be made to interfere with the Althing's proceedings by violence. The police force was small, even though it had been augmented somewhat by volunteers. The leaders of the three Government parties thus issued an appeal to peace-loving citizens to come to the House "in order to demonstrate by so doing that they want the Althing to have peace to carry out its duties."

A large crowd gathered outside the House. Inside, members argued in vehement debate. Outside, there were soon scuffles and shouting, and stones were thrown at the parliament building, breaking windows in the Chamber. The situation soon became intolerable. The police, together with its reserves, began to clear the area around the House using both truncheons and tear-gas.

The Althing got the peace it needed to conclude the debate. Amendments, proposing that Iceland's unique position be recognised in the text of the Treaty, calling for the termination of the 1946 Keflavik Airport Agreement and demanding a plebiscite on membership of the Alliance, were all defeated. The voting went 37–13 in favour of joining NATO. Apart from the Communists, two Social Democrats and one Progressive member voted against the resolution and two other Progressives abstained. The Government supporters who opposed the motion had all voted for the defeated amendments. Some of those who voted in favour were subjected to rough handling and abuse when they left the House on that historic day. But as time went by the storm died down, and at least two of the three Government supporters who had voted against the Resolution were later to announce that they considered that the decision to join the Alliance had been the right one.

The address which Foreign Minister Bjarni Benediktsson delivered when he signed the North Atlantic Treaty on 4 April 1949 in Washington began as follows:

"The nations who are now forming this new alliance are unlike each other in many respects. Some of them are the greatest and most powerful in the world. Others are small and weak.

"None is smaller or weaker than my own — the Icelandic nation. My people are unarmed and have been unarmed since the days of our Viking forefathers. We neither have nor can have an army. My country has never waged war on any country, and as an unarmed country we neither can nor will declare war against any nation, as we stated when entering the United Nations. In truth, we are quite unable to defend ourselves from any foreign armed attack.

"There was, therefore, hesitation in our minds as to whether there was a place for us as participants in this defensive pact. But our country is, under certain circumstances, of vital importance for the safety of the North Atlantic area. In the last war, Great Britain took over the defence of Iceland, and later we concluded an agreement with the United States Government for the military protection of Iceland during the war. Our participation in this Alliance shows that for our own sake, as well as for the sake of others, we want similar arrangements in case of a new war, which we all indeed hope and pray never will occur."

The world security situation unfortunately did not improve over the next few years and in the spring of 1951 Iceland thought it best to conclude a bilateral agreement with the United States, under the auspices of NATO, on the stationing of American troops in the country. This agreement is still in force, and from that day to this Iceland's security has rested, on the one hand, on the North Atlantic Treaty and, on the other, on this bilateral agreement.

The great political battle, some of the main features of which have been described in this article, led the Icelanders to conclude that the independence they had so recently regained was, in times of uncertainty, best safeguarded in free co-operation with those peace-loving democratic neighbours with which they have the closest cultural and ideological links. The hopes associated with this co-operation have not proved ill-founded, and this is still in the estimation of the great majority of the nation — the most sensible option.

AN ALLIANCE CLAMOURING TO BE BORN — ANXIOUS TO SURVIVE

André de Staercke

The North Atlantic Treaty Organisation was born out of fear and hope: the instinct for self-preservation in nations emerging from the war and afraid of losing their freedom was triggered by fear; hope sprang from their union, opening up a vista of a new unknown landscape still being explored more than thirty-five years later.

The origins of the union can be traced to the concluding days of the war with Japan, when on 12 May 1945, a few days after Germany's capitulation, Winston Churchill in a prophetic telegram warned President Truman: "I am deeply concerned by the Russians' misreading of the Yalta agreements . . . . What will be the position in a year or two when the British and American armies have melted, and the French have not yet been formed on any major scale, and when Russia may choose to keep 200 or 300 divisions on active service? An iron curtain is drawn down upon their front . . . ."

What did in fact happen during those two years? Eastern Europe, an area of 1,400,000 square kilometres with approximately 87 million inhabitants, fell under Soviet domination in a "conquest without war". The tide of Russian expansionism advanced towards Greece, threatening the annexation of Turkey, the Straits and the Middle East. On 12 March 1947, just less than two years after Churchill's warning, President Truman gave his reply before the United States Congress: "It must be the policy of the United States of America to support free peoples who are resisting attempted subjugation by armed minorities or by external pressure." This was the first enunciation of the Truman Doctrine, a policy which was to have an immediate and decisive effect on events.

It was the Truman Doctrine which was to form the basis for structuring the recovery of the Western world and providing for its future defence. The United States Congress began by authorising 400 million dollars' worth of aid to Greece and

Turkey. Two months later, on 5 June 1947, General George C. Marshall, then Secretary of State, launched the European aid plan which was to bear his name.

But Soviet hunger for expansion continued unabated. On 22 February 1948, in the first Prague coup, Russia gained control of Czechoslovakia. Time was pressing. On 17 March 1948, Belgium, France, Luxembourg, the Netherlands and the United Kingdom signed the Treaty of Brussels, which promised a system of common defence. Although the measures envisaged in the Treaty were themselves laughably inadequate, there were nevertheless two immediate reactions: one of fundamental importance on the part of President Truman, within the framework of his Doctrine, and the other no less important — and alarming — on the part of the Soviet Union: the blockade of Berlin.

On the very day the Treaty of Brussels was signed, President Truman declared to Congress that the resolve of the free countries of Europe to defend themselves must be equalled by the resolve of the United States to assist them. Work began immediately. The Berlin blockade was thwarted without bloodshed by a plan as bold as it was unique: an airlift, which was to last for 323 days. At the same time, it was decided to prepare the ground for a common defence system, building on but extending well beyond that provided for by the Treaty of Brussels. On 29 April 1948, the Canadian Secretary of State for External Affairs, Louis St Laurent, put forward a proposal to the House of Commons in Ottawa, and Senator Vandenberg managed to get the American Senate by a large majority — 64 votes to 4 — to adopt the famous resolution that bears his name, proposing that a common defence pact be negotiated.

On 9 July 1948 discussions began in Washington. On 28 September 1948 Paul-Henri Spaak, at the United Nations General Assembly in Paris, made his famous speech on fear, which met with a triumphant reception. Upbraiding Mr Vyshinsky, the Soviet Representative at the UN Assembly, he declared that the only explanation to be sought for the reactions of the Western nations was fear — fear of the USSR, fear of its government, fear of its politics. On 15 March 1949 Denmark, Iceland, Italy, Norway and Portugal were invited to join the United States, Canada, Belgium, France, Luxembourg, the

Netherlands and the United Kingdom in the new Treaty Organisation — twelve nations in all. Three weeks later, on 4 April 1949, despite last-minute manœuvres and threats from Moscow, the North Atlantic Treaty was signed in Washington. One month later, on 9 May 1949, Stalin lifted the Berlin blockade, nearly a year after it had begun. Less than four years after the end of the Second World War, the Truman Doctrine, fear, courage and the will to live had given the West an Alliance which was to provide lasting peace, and which offered, as indeed it still does, untold possibilities for the future.

This Alliance, by bringing together sovereign nations so different in size, so far apart, so diverse in their origins, offered history a short cut which would have been ignored had not the pressures been so great. The Allies' commitment was without precedent, save perhaps for the Confederacy of Delos or Delian League (for the Greeks invented everything) which, in the fifth century B.C., in a time of peace and under the patronage of Athens, established a confederation of Greek states with a common treasury to finance individual contributions towards their common defence against the threat from Persia. The League lasted for sixteen years, ending in bickering and indifference. But let's not take our analogy too far: I for my part can think of no other democratic alliance that has ever succeeded, in times of peace and for an indefinite period, in building anything approaching a common political and military policy.

The first three years of the Alliance were difficult, as they were bound to be. Everything was new. The object was clear: not to win a war but to prevent peace from being lost. The search for the means led to confusion, described in masterly fashion by Eve Curie in the report which she wrote for Lord Ismay under the title "NATO, the First Five Years". I shall not, therefore, dwell on the succession of ministerial conferences, the rivalry between mutually independent ministerial committees, military committees, civilian committees, the duplication of jobs and conflicts of jurisdiction. It was all a flurry of rushed improvisation, in the shadow of imminent danger. This was the time when Dean Acheson, the American Secretary of State, declared to one meeting of foreign ministers that the USSR was a universal conspiracy putting itself forward in the guise of a state. To face up to this conspiracy, it was the military bulwarks that were the first focus of attention. In a deluge of ideas and decisions, a military committee representing the member-countries was called upon to meet in Washington. Its executive body, the Standing Group, comprised the "Big Three" — Britain, France and the United States. As ministerial meetings succeeded each other, new ideas emerged and a strategic concept was defined based on deterrence, the acceptance of integrated defence, the establishment of a four-year medium-term defence plan, balanced collective forces, and equitable distribution of the defence burden — all concepts which, with others, continue to sustain the ideology and vocabulary of the North Atlantic Treaty Organisation.

After a number of tentative efforts, however, it was realised that leadership of undisputed reputation was needed. Given the panic caused throughout the West by the Korean war, rather like the terror of the year 1000, the need for the Alliance had become all the more acute. General Eisenhower was called upon. He hesitated, since the destitution of the free world was such that he wanted first to be certain of its will and its ability to defend itself. When, in April 1951, he agreed to become Supreme Commander in Europe, he called to his side his former comrades-in-arms and in the front line Field-Marshal Montgomery, who brought with him a spectre: the military organisation of the Treaty of Brussels. Alongside this prestigious constellation of military figures, the governments of the North Atlantic Treaty Organisation had, in May 1950, created a Council of Ministerial Deputies. Through the chaos of those early days, it was the task of this new body to set about building, systematically and progressively, an organised political and military structure for the Alliance, and to promote the organisation of its forces.

Since the Military Committee was located in the United States and the Supreme Command (SHAPE) in France, concern for political balance dictated that the Council of Deputies naturally had to be based in London. This did not help to ensure rapid and frequent contact between the three. The Council Deputies made their headquarters at Number 13, Belgrave Square. In the face of the formidable military organisation which was being developed at the time over the entire combined territories of the Alliance, and in the face of the authority of the

Supreme Commander General Eisenhower, who had directed the victorious armies in the Second World War and was about to become President of the United States, the Council of Deputies carried little weight. Its terms of reference were vague; nor was it always certain which Ministers it represented. To the international and even to national military authorities, it seemed a pointless distraction in their relations with the Ministries of Defence. When things were going well, it was ignored. When things were going badly, it was blamed — rather as today, after thirty-five years of existence, the inadequacies or failings of individual Allies tend to be attributed to a failure on the part of the Alliance and of its Council. Left to rely on its own resources, the Council of Deputies had no real administrative services other than those provided by each Deputy, which in many cases did not amount to much. Some countries were represented by their Ambassadors in London, others by civil servants sent from their capital cities. This was my own case, as the Belgian representative. I divided my time between Brussels, London and Paris, where I was also responsible for directing the Belgian delegation at the Conference for the creation of a European army. This situation merely reflected the rush and merciless pressure of events.

Yet the Council of Deputies accomplished a great deal. Despite the inadequacy of its resources, or maybe because of it, those who belonged to the Council had a faith and enthusiasm which eventually was to make their efforts worthwhile. They worked under the Chairmanship of the American Deputy, Charles Spofford, a conservative legal specialist, untiring, obdurate, but above all blessed with vision. I should really mention them all individually, but failing that I would at least like to offer them, collectively, the praise which they deserve. If the North Atlantic Alliance, this great adventure, has become what it has, it is because of the value of their contribution to its early history. From the very beginning they understood that the scant treaty of 14 Articles meant an opportunity. Its very simplicity opened up an uncluttered view of the future, over and beyond present vicissitudes. Very quickly too they realised that if the Alliance was to last, it must have a life and form of its own: it must have a sufficiently distinct identity to enable it to act in the interests of its member-states without becoming confused with each or any of them individually. In other words, it must become more than the sum of its parts. They also saw that the Alliance had to become something other than a military structure for, as the Bible has it, people tire of armies, however much they need them. So the structure had to have a powerful civilian organisation, to direct and support it — through the cold war and the continued defence effort — towards the relaxation of tensions which was supposed to follow.

With their governments behind them, the Deputies worked towards these goals, and since it was one of those new eras when living meant creating and not merely sticking to existing arrangements, plans for the future were tackled alongside problems requiring immediate solution. Two events of fundamental importance stood out on the immediate horizon. The first was the protocol of accession of Greece and Turkey to the North Atlantic Treaty, signed by the Deputies on 22 October 1951. Without Greece and Turkey, which were already participating in planning for the defence of the Mediterranean region, the protection of the southern sector would have been incomplete, perhaps even impossible — a fact recognised after protracted consultations. If originally there was some evasiveness in various quarters, this disappeared in the feeling experienced on all sides of sharing in a common destiny, and the hope that the two new Allies, historically adversaries but naturally destined to understand one another, would be guided first and foremost by the external threat which was the true concern of the Alliance, rather than dragging the latter into taking sides and finding solutions to their secular disputes. It was a challenge for the future.

The Deputies had to deal with another urgent problem: the cost of defence. Here there was almost total lack of co-ordination. The military authorities called for contributions which the economic situation of the Allies could not sustain. There was a risk that the warrior would be crushed by his own armour. The effort involved in surviving threatened to destroy the survivors. Urgent action was therefore needed. The French Deputy, Hervé Alphand, invoked Article 2 of the Treaty to recommend controls on the measures not only required for the security of the Atlantic world but also essential for its economic stability.

With this in mind, at the Ottawa Ministerial meeting of

1951* the Council set up a Temporary Council Committee (TCC) with the widest powers of investigation and proposal. The aim was to reconcile the requirements of defence with the economic capabilities of the Allies. Its members acted not as representatives of their countries, but as representatives of NATO. Its Executive comprised three members. To further affirm their independence, three leading "wise men" were chosen: Averell Harriman (United States), Jean Monnet (France) and Sir Edwin Plowden (Great Britain). Under the stimulation of these three, the TCC acted like a whirlwind and in December 1951, three months after its creation, it presented its report. General Eisenhower described it as a "truly monumental work". Indeed the conclusions which could be drawn from it dissipated the incoherences and contradictions of Atlantic rearmament. They allowed the Deputies to suggest an impartial process for planning their combined effort to which the Allies inclined: assessment of the threat, identification of requirements, and proportional contributions to the military goals of the Alliance related to the economic and social capabilities of the member-states and established in the form of commitments worked out on the basis of mutual examination. This was the beginning of an "annual review" procedure which, with necessary adjustments, still applies.

So far I have endeavoured, through a flurry of dates and problems, to indicate the breathless pace of the first three years of the Alliance. It finally emerged from its initial confusion, and the Deputies were able to step back, in accordance with their Governments' instructions, and begin to plan for a future which, if no more tranquil, was at least better organised. From the many potentiatities of the Treaty, they conceived a structure which was adopted at the Lisbon Conference in February 1952. This provided for an indivisible Atlantic Council, with full powers, irrespective of the level of its meetings, whether Heads of State or Government, Ministers, or permanent representatives. It created the position of a Secretary General, who was to

---

*As a measure of the wide-ranging coordination which was then taking place, for the first time Ministers of Defence and Economics or Finance participated in the Ottawa meeting of the Council, along with their Foreign Ministry colleagues.

preside over the Atlantic Council at all levels, assisted by an International Secretariat made up of all the civilian bodies in the Alliance. The Lisbon reform was fundamental, and thanks to it the Alliance moved from improvised arrangements to institutional forms. It nevertheless retained, as it still does, the operating flexibility that had characterised its birth. It is my profound belief that if it manages to preserve for itself this status, with its imprecision — or rather, progressive precision — in which past practices and precedents evolve smoothly from the needs of the present to future requirements, it will remain an incomparable instrument in which member-states can demonstrate their solidarity without sacrificing their sovereignty.

It is this aspect that allows me to speak of the Alliance's triumphs, the first of which was the development of the function of Secretary General. The nature of things, and exceptional personalities, have made the Secretary General what he is today, the embodiment of the Alliance. In 1956, just after the Suez crisis, three new wise men, Halvard Lange of Norway, Gaetano Martino of Italy and Lester Pearson of Canada, in their famous report to the Council on non-military co-operation within the Alliance — a report now too frequently forgotten — defined the substance of the Secretary General's powers without setting limits to them. The member countries' respect has done the rest, with the constant support of the Council and the sense of balance shown by each of the Secretaries General. Once again, I believe that the effectiveness of the Alliance is linked to its resolute support for the individual who is the personification of the institution.

Since I am speaking here of the beginnings of the Alliance, I shall refer only to the first two Secretaries General, both of whom greatly influenced its evolution. "Go to that Council of yours", Winston Churchill said to me one day in March 1952, "and tell it that I'm giving it my right arm, Lord Ismay. I could make no better gift, give no better man. The rest will follow." And the rest did follow. Former General in the Indian Army, one of Churchill's most intimate colleagues during the war, advising him on military matters, and a member of the British Government, he was the ideal man to stimulate the unified activity of the Alliance following the Lisbon reform. For those who did not know him, it is not obvious how much NATO owes to him. Behind his good-natured exterior was a keen eye, a firmness, a resolve and immense experience. He it was, in affable and serious mode, who established the habit of continuous dialogue between governments that the Atlantic Council represents. He it was who achieved the *tour de force* of obtaining a gentlemen's agreement from the Allies on the use of English and French as the official languages of the Alliance, to avoid a tower of Babel of different tongues. He it was again who pursued the difficult and still unfinished course of NATO's military organisation — difficult because in the aftermath of war, faced with an imminent threat, just as the Roman Republic placed its fate in the hands of a temporary dictator, the free world entrusted its survival to the eminent military leaders who had recently saved it. This situation could only be temporary, since it was liable to jeopardise the entire democratic system. It is to the military authorities' credit, and to that of Lord Ismay in particular, supported by General Eisenhower, that they understood that NATO's success could only be achieved and could only be durable if guaranteed by a political authority. Under the impetus of the former General Ismay, the primacy of civilian power was imposed over the military structure without restricting its operation.

Lord Ismay opted to leave in 1957 after five years in an office on which no time limit had been imposed. "I have been the wet nurse," he said; "now it is time for the governess." This was to be Paul-Henri Spaak, Belgian Minister of Foreign Affairs, First President of the General Assembly of the United Nations in San Francisco, and one of the fathers of Europe. He was, to quote Plato, "a great man, born in a little village, and as such attentive to world affairs". In Spaak's hands the Council became an efficient instrument of political consultation. Beyond the military protection of the Alliance, and thanks to that protection, peaceful coexistence, which the free world was later to term détente, began to emerge from the cold war atmosphere. Spaak devoted himself wholeheartedly to cultivating it. So the marriage of defence and détente was born, gradually to become the pivot on which the Alliance's policy turned in one direction or another, according to the vicissitudes of the balance of forces and the fluctuations of the external threat. It was another Belgian Minister of Foreign Affairs, Pierre Harmel, who was to proffer a masterly definition of this policy in 1967, in a striking report on the future tasks of the Alliance which bore his name and which spelled out the hopes and possibilities of future negotiations, particularly for disarmament.

But Spaak was looking much further ahead. He was an advocate of the development of NATO in the direction of a two-pillared Atlantic Community — the American pillar and the European pillar, on the lines favoured by President Kennedy. In pursuit of the logic of this concept, which claimed progressively increasing solidarity from the Allies, the precision of his own thinking induced him to recommend a form of consultation which even extended to some harmonisation of policies in certain spheres outside the geographic perimeter of the Alliance. He was not followed, for reasons relating to the sovereignty of each state, and doubtless also because he was going too fast too soon. He resigned, and left office on 31 March 1961, returning one month later to his position as Belgian Minister of Foreign Affairs. The letter he wrote to President Kennedy explaining the reasons for his resignation, and reproduced in his memoirs, has lost none of its relevance and repays re-reading.

Another success of the Alliance has been the way in which the Atlantic Council has evolved. Under the Lisbon Reform this was a body meeting intermittently at ministerial level, but in continuous session at permanent representative level, based first in Paris and then in Brussels. In an empirical progress which was not without accidents, the Council, particularly at permanent level, became to some extent the repository of Western security. As such it became the centre of gravity from which the many and various wills of the Allies gain the strength of collective action. And as such it is still the place where their differing policies are clarified and coordinated, and even sometimes where unity is achieved. Its composition makes it an ideal meeting place, a venue for frank confrontation between North America and Europe. Finally, the extreme flexibility of its activities and lack of rigidity of its procedures enable it, where necessary, to adjust relationships between Allies without jeopardising the Alliance itself. The Permanent Council has thus become the "decent club" to which Lord Ismay referred, "one of the only places in the world", he used to say, "where friends can fight and even agree"

"Where friends can fight and even agree"! These words

could well be inscribed, among others, over the entrance to the Council chamber. From further back than I can remember, which means since the beginnings of the Alliance, there has been a continuous series of crises — but within the democratic context of NATO they are its very life and not crises of the Alliance itself. Let us look at a few examples. Since 1950, encouraged by the Americans, NATO began to envisage German participation in Western European defence. The war had only recently ended, and its memories were still too immediate for the suggestion not to arouse sharp controversy. But the notion of Europe had been born. The United States encouraged it, and General Eisenhower nurtured a truly European ideal. The Schuman plan, setting up a coal and steel community, had just got under way. Chancellor Adenauer offered his cooperation. France then proposed, under the Pleven plan, that the future German military contribution should be integrated in a European army which could be put at the disposal of the Alliance. NATO agreed, and the conference on the creation of a European army opened in Paris in February 1951. For three years we slaved away enthusiastically, and we felt that we were ploughing the fields of the future. In 1954 we were on the point of succeeding. The treaty setting up a European Defence Community (EDC) was ready for signature when the insurmountable political obstacle arose. I had expected it. One day, at Chartwell, towards the end of our negotiations, I had told Sir Winston Churchill of my apprehensions. I stressed, in particular, that in the system which we were proposing "the French were too much inside and the British too much outside," that because the military function was of prime importance in the life of states, the sharing of it could only be contemplated if there were no exceptions.

The future cannot be hurried. The idea was a wonderful one, and its implementation could have provided many solutions to problems that are with us today, but it was doubtless premature. Nevertheless, its failure was a disaster for Europe, and a setback for the Alliance. The latter found the means of its salvation in its own vitality. Sir Anthony Eden immediately took up the pilgrim's staff to find an alternative solution. Two months after the failure of the EDC, on 23 October 1954, the Paris agreements were signed. They invited the Federal Republic of Germany to join NATO and they set up the Western European Union (the

WEU), adding West Germany and Italy to the five countries of the Treaty of Brussels. West Germany thus became the fifteenth member of the North Atlantic Treaty Organisation. A crisis within NATO, which could have become a crisis for NATO, had been resolved.

The Suez affair in 1956 was another crisis — and what a crisis! It was all the more unfortunate in that the Franco-British initiative against Nasser almost coincided with the USSR's march into Hungary. The Alliance was weakened by a military intervention outside the Treaty area delivered, like a bolt out of the blue, by two of its leading members. It became increasingly clear that political consultation between Allies had little impact on actual intentions. It was also clear that Atlantic solidarity was incapable of surviving in a system where member countries were friends in one part of the world and adversaries in another. The Alliance was in total disarray, and it needed nothing less than the three wise men (Lange, Martino and Pearson) to sort it out. They considered what were decorously called the "serious tensions" of the relations between Allies, and in a matter of a few weeks produced a remarkable report on political co-operation, which the Council of Ministers approved in December 1956. NATO was extricated from the morass, but the problem of the direction and scope of consultation was unresolved. It remained and still remains the key to the success of the Alliance in the future.

The crisis that erupted in 1966, when France decided to withdraw from the military structure of the Alliance, could have been fatal. The entire edifice of NATO appeared to be in jeopardy. Would it collapse? Once again the danger was averted, thanks to negotiations in the course of which goodwill on all sides produced the optimum results. Still part of the Alliance, France remained a full member of the Atlantic Council governing the general policy of the Alliance, while her fourteen partners, with France's agreement, founded a Defence Committee in which they continued to develop the military organisation. As doyen of the Permanent Council at that time, I had the honour of presiding over the team of fourteen negotiating with France, and I admit that as a result of the solution obtained, I have never regarded France's withdrawal from integration as a catastrophe. On the contrary, from that time the fourteen were able to pursue, in unison, the essential task of setting up integrated forces; they were able in particular, without difficulty or contradiction, to adopt the strategic concept of flexible response, with all its consequences for the notion of deterrence and the structure of the forces it required. For her part, France remained the faithful ally which she had always been, and has never ceased to be, while at the same time remaining both in control of her own military options and independent within the context of the ties binding the Alliance.

What conclusions can be drawn from the ups and downs that have marked the internal progress of NATO since it first came into being? The first is the growth of a formidable collective vitality, which has so far managed to overcome the cumulative obstacles to its effectiveness. Then a far from negligible result: peace. Not universal peace, since many conflicts still exist and will continue to exist in the world for a long time to come, but peace within a vast region where a flare-up between two giant antagonists could threaten the survival of the planet. By blocking the expansion of Soviet Communism in the West, the Atlantic Alliance has rendered a worthy service to humanity, and those who have contributed to this great enterprise may feel legitimate pride. No apologies are needed for having safeguarded peace.

It is, however, a continuing task. Who will now take it over, and how? I recall the passage in the *Acts of the Apostles* when the feet of the young men are heard echoing at the door, coming to take the spoils of the old. How will the youth of the free world manage this peace, bequeathed by earlier generations and now in its own hands? By unilateral pacifism? There is little point in sheep supporting vegetarian notions if their views are not shared by the wolf. Here President Mitterrand's words are precise and to the point: "We have pacifism here in the West, but missiles in the East." This form of pacifism seems to me tied to an escapist philosophy which refuses to look deeper into the problem to avoid having to solve it. What choice will they make, those who now hold their future in their hands? Shall it be pacifism generated by fear, with nothing to bring to the negotiating table but the determination to assert itself? Or security based on a balance of forces with its real promise of fruitful negotiations?

There are other tests that lie in wait for the Atlantic Alliance, as it enters the latter half of the 1980s. The first is what I would call the habit of peace. It offers a great temptation: the longer we live in peace, the less aware we become of the danger. We find it difficult to imagine that the long-averted threat still exists, if indeed it ever did exist. It takes a great deal of wisdom to pursue an effort when the very success of the undertaking is its own worst enemy. And to this test must be added another: the impact of the present and its inevitable constraints, and the risk that it may cloud our vision of the future. We should indeed be concerned by the seeming compulsion to treat the present more or less in isolation, as a reflection of the sovereignty of the member-states, their national interests or their internal difficulties. This is a form of creeping unilateralism, which would result in a nominal alliance in terms of objectives but one divided in practice in terms of actions. It would be a rout by another name — which is why the reality of the consultation process and the pursuit of progress in the manner of its realisation are more necessary than ever. It is also a great discipline to know how to explain ourselves to one another, and to be jointly associated in striving for the same objective, without supremacy of the greater or neglect of the weaker members. This is the price of the cohesion of the Alliance, so difficult to preserve.

The drifts created by these various trends are by no means necessarily fatal; the Alliance has seen them before and will see many others like them; but it is doubtless these which explain the controversies, speeches and proposals which abound today in an effort to correct its course. The conclusion that has to be drawn is an optimistic one, since we would hardly allow it so much attention if we did not regard it as essential.

It is on this note of hope that I should like to end. The Atlantic Alliance embodies the will of the West to live and to survive. Its sixteen members (for Spain has now joined — a further milestone which we hope will be a lasting arrangement), through their membership of the Alliance, cast their vote for a civilisation based on liberty and democracy. It threatens no-one, and as such can be supported and respected by all countries in the world which believe that their destiny is their own and must not be determined by a foreign power. If the military organisation of the Alliance is essential to ward off the prospect of conflict, its response to the challenge which confronts it remains fundamentally political. Its object is peace in security, to be achieved by untiring negotiations and reciprocal disarmament. With the development of its many other activities, this remains for the Alliance, thirty-five years on, the permanent task of the future. The promise of the Angel of the Apocalypse is not inappropriate: "I have opened before you a door which no-one can close."

PRINCIPAL EVENTS, CONFERENCES AND TREATIES

The following historical notes describe the principal events, international conferences and treaties referred to in this volume. The arrangement is chronological.

Treaty of Versailles

Signed on *28 June 1919*, between the Allied Powers and Germany, the Treaty signalled the end of the First World War. It provided for the establishment of the League of Nations; the return of Alsace and Lorraine to France; the ceding of other territories to Belgium, Denmark, Poland, Czechoslovakia and Lithuania; and the demilitarisation of the Rhineland. The Treaty was ignored by Hitler in the years preceding the Second World War.

Foundation of the League of Nations

The League of Nations was the first attempt to create a worldwide organisation to prevent war and encourage international cooperation. Established in Geneva on *10 January 1920* on the basis of the agreement reached at the Treaty of Versailles, the League had 42 member-nations. The United States, whose Congress failed to ratify the League Covenant because it considered that its conditions provided insufficient safeguards for American sovereignty, did not join. Effective between 1920 and 1932 in arbitrating in a number of European territorial disputes, the League's authority declined in the 1930s when it failed to check the expansionist policies of powerful and increasingly totalitarian regimes such as Japan and Nazi Germany. The Soviet Union joined the League in 1934, but was expelled following its attack on Finland in December 1939. The League proved ineffectual in protecting Abyssinia (Ethiopia) after its invasion by Italy in 1935, and failed to prevent aggression elsewhere. It continued to exist in name only until 1946 when its functions and institutions were transferred to the newly-established United Nations Organisation.

Treaty of Svalbard

Signed in Paris on *9 February 1920*, the Treaty granted sovereignty of the Svalbard archipelago (including Spitsbergen and other islands) to Norway and mineral rights to Denmark, France, Italy, Japan, the Netherlands, Norway, Sweden, the United Kingdom, the United States and, in 1925, the Soviet Union. Today only Norway and the Soviet Union continue to exercise these rights for the mining and exporting of coal. The archipelago has been used for a number of scientific expeditions and remains of strategic importance to Norway and to NATO's northern flank.

Principal Events, Conferences and Treaties

Formation of the Rome-Berlin Axis

The name given to the union of Nazi Germany and Fascist Italy formed in *October 1936* to ward off sanctions imposed on Italy by the League of Nations, following its invasion of Abyssinia. The union became a full military and political alliance in May 1939. It was extended to include Japan in September 1940 and subsequently took in Hungary, Bulgaria, Romania, Slovakia and Croatia. The Axis collapsed with the fall of Mussolini and Italy's surrender in 1943.

Yalta Conference

Wartime meeting in *February 1945* between Winston Churchill, Prime Minister of Great Britain, President Roosevelt of the United States and Joseph Stalin, Premier of the Soviet Union. Often incorrectly regarded as the event which legalised the division of East and West Europe, the Yalta Conference laid the ground for the post-war administration of Germany and the territories under German occupation. It took place at a time when Britain, the United States and the Soviet Union were allied against an all but defeated Germany, and when some in the West still entertained illusions that this wartime alliance would be translated into peacetime cooperation. It was in this optic that agreement was obtained on a charter for the future United Nations Organisation.

Specific agreement was also reached at Yalta on a number of more immediate concerns, namely the division of Germany into four zones of occupation; the complete disarmament of Germany; the reestablishment of Poland with new frontiers; the declaration of war on Japan by the Soviet Union and the post-war cession to it of disputed Japanese islands; the reestablishment of the independence of Korea after a predetermined period of joint US/Soviet military occupation; the endorsement of the "Atlantic Charter" announced by Churchill and Roosevelt in August 1941 as the basis for Allied post-war policy; and the establishment of democratic governments in all liberated countries on the basis of free elections.

Signing of the United Nations Charter

The Charter of the United Nations was signed by 51 nations on *26 June 1945* following a series of conferences held to discuss the formation of a supranational organisation to succeed the failed League of Nations. The first of these conferences was the Moscow Foreign Ministers' Conference of October 1943. This was followed in August 1944 by the Dunbarton Oaks Conference in Washington DC attended by Britain, China, the United States and the Soviet Union. At the Yalta Conference of February 1945, approval of the United Nations Charter was finally obtained, but at the insistence of the Soviet Union it was made subject to insertion of the controversial right of veto, allowing a dissenting member of the proposed UN Security Council to defeat majority decisions.

Article 51 of the UN Charter, which establishes the inherent right of member-nations to individual or collective self-defence in the event of armed attack, is the basis for the North Atlantic Treaty of 1949. The United Nations works through its six main organs: the General Assembly, the Security Council, the Secretariat, the Economic and Social Council, the Trusteeship Council and the International Court of Justice. The last two of these had been created as instruments of the League of Nations, but were transferred to the United Nations in April 1946. The United Nations also operates through a number of specialised agencies such as the International Labour Organisation (ILO), the General Agreement on Tariffs and Trade (GATT), the UN Educational, Scientific and Cultural Organisation (UNESCO) and the Food and Agricultural Organisation (FAO).

Potsdam Conference

Attended by President Truman of the United States, by Joseph Stalin, Premier of the Soviet Union, and by the Prime Minister of Great Britain (Winston Churchill and, after the election of a Labour Government in Britain, Clement Attlee), the Potsdam Conference took place in *July-August 1945* after the surrender of Germany at the end of the Second World War. It resulted in agreement on the need for the unconditional surrender of Japan, and constituted the basis for the post-war administration of Germany by Allied forces. Detailed decisions were taken on reparations, the dismantling of the German industrial complex, the demilitarisation of Germany, the decentralisation of its political institutions, and the transfer of certain territories to Poland and to the Soviet Union.

Treaty of Dunkirk

Signed on *4 March 1947* by France and the United Kingdom as a 50-year mutual defence pact against a revival of German aggression. Proposals made by Ernest Bevin, Foreign Secretary of the United Kingdom, in January 1948 cited the Dunkirk Treaty as a model for a network of bilateral treaties which together would constitute the Western Union.

The Dunkirk Treaty differs from the North Atlantic Treaty particularly in that it is of 50 years' duration and was aimed expressly at preventing renewed aggression by Germany. The North Atlantic Treaty is of unlimited duration and is designed to counter aggression against the members of the Alliance from any external source.

Truman Doctrine

Described in the first chapter of this book, the Truman Doctrine enunciated by the United States President on *12 March 1947* pledged American support for 'free peoples who are resisting attempted subjugation by armed minorities or by outside pressures'. The immediate impetus for this policy was Britain's inability to provide the economic and military aid necessary to assist Greece and Turkey to resist a Communist takeover. The US Congress subsequently agreed to appropriate $400 million for this purpose.

The significance of the Truman doctrine lies in the role it played in reversing traditional American isolationism. It provided the conceptual framework within which concrete programmes of economic and military assistance were subsequently elaborated. It therefore played a decisive role in containing Soviet expansionism while establishing the basis for the post-war transatlantic cooperation on which the North Atlantic Alliance was founded.

Brussels Treaty of Economic, Social and Cultural Collaboration and Collective Self-Defence

Signed in *March 1948* by Belgium, France, Luxembourg, the Netherlands and the United Kingdom, the Brussels Treaty represented the first formal step in the direction of the North Atlantic Treaty by establishing what became known as the Western Union. It emerged as the first multilateral regional arrangement for the security of Western Europe to be established under the United Nations Charter, in preference to the bilateral treaties (on the lines of the Dunkirk Treaty) which had first been proposed. The immediate effect of the Brussels Treaty was to provide the US Administration with the evidence needed to convince Congress of Europe's determination to organise itself for defence – a prerequisite for American involvement in the future Atlantic Alliance. The Brussels Treaty powers established a political structure, consisting of a Consultative Council of Foreign Ministers and a Permanent Commission, and a Military Committee. Each of these played a significant role in the further negotiations leading to the North Atlantic Treaty. The Military Committee of the Brussels Treaty powers was in fact created in response to an American request for information on European military plans and actual and potential sources of military supplies, to assist in preparing Congress and US public opinion for American participation in the North Atlantic Treaty. Like the Dunkirk Treaty, the Brussels Treaty was limited to a period of fifty years.

In September 1948 a military body was created within the framework of the Brussels Treaty, known as the Western Union Defence Organisation. Field-Marshal Montgomery was appointed Chairman of the Commanders-in-Chief and set up his headquarters at Fontainebleau. Commanders-in-Chief were named for Land, Air and Naval forces. Following the signing of the North Atlantic Treaty in April 1949, the Brussels Treaty powers agreed to merge the military structure of the Western Union with NATO. The responsibilities of the Western Union Commanders-in-Chief Committee were transferred to General Eisenhower in April 1951 when he became NATO's first Supreme Allied Commander Europe, and the staff and facilities of the land, sea and air commands of the Western Union were placed at his disposal. Under the Paris Agreements of 1954, the Federal Republic of Germany and Italy acceded to the Brussels Treaty and the Western Union became the Western European Union (WEU).

European Recovery Programme (Marshall Plan)

With Europe facing the prospect of post-war economic collapse, George C. Marshall, US Secretary of State, initiated the idea of a major Programme for European Recovery. His proposal, outlined in a speech given at Harvard University on *5 June 1947*, amounted to an offer of economic assistance extended to the whole of Europe. The offer, 'directed not against country or doctrine but against hunger, poverty, desperation and chaos', was rejected outright by the Soviet Union which saw it as an instrument of American imperialism. The plan was warmly accepted by Britain, France and fourteen other West European countries. Following the Paris conference on European Economic Cooperation of July 1947 the proposal was translated into legislation and gave rise to an extensive programme of economic assistance and reconstruction, and to a new optimism which was to pave the way for subsequent efforts to establish economic cooperation within Europe.

The programme, amounting to some US$13 billion paid between 1947 and 1951, laid the foundations of European economic stability by stimulating production, increasing European exports to North America and overcoming the problem of European dollar deficits. Its management was entrusted to the Organisation for European Economic Cooperation (OEEC), later to become the Organisation for Cooperation and Development in Europe (OCDE).

Treaty of Rio de Janeiro

Signed on *2 September 1947* by the United States and 20 countries of Latin America (excluding Nicaragua and Ecuador), the Rio Treaty is an Inter-American Treaty of Reciprocal Assistance binding the signatories to support its members in the event of aggression against them. It established a collective defensive alliance within the framework of the United Nations Charter. Unlike the 1949 North Atlantic Treaty, for which it was one of the models, the text of the Treaty makes no reference to the use of armed force in response to aggression, but provides for joint action to deal with emergencies.

Warsaw Conference of Communist Parties – Establishment of Cominform (Conference of Sklarska-Boreba)

An abbreviation for the Communist Information Bureau, the Cominform was established at the Warsaw Conference of the Communist parties of the Soviet Union, Poland, Bulgaria, Rumania, Yugoslavia, Czechoslovakia, Hungary, Italy and France, held between *22 September and 5 October 1947*. Yugoslavia was expelled from the Cominform in June 1948, which was finally dissolved in April 1956. The aim of the Cominform was to coordinate the political activities of the Communist movement in European countries. It was seen as the Soviet Union's response both to the Truman Doctrine and to the Marshall Plan for providing American aid to European countries whose economies had been devastated by the Second World War. The Soviet Union vetoed Poland and Czechoslovakia's participation in the Marshall Plan, and continued to regard Marshall Aid as 'an instrument of American imperialism' rather than a programme for the economic recovery of Europe. In 1949 it established the Council for Mutual Economic Assistance (CMEA) as a more direct response. These developments reflected the futility of initial post-war hopes of constructive East-West cooperation and accentuated the division of East and West into two camps.

The Cominform should not be confused with the "Comintern" (Communist International) established in March 1919 to promote the proletarian revolution throughout Europe. The latter became a vehicle for suppressing internal opposition to Stalin in the 1920s, but by 1930 its function was limited to organising support for Soviet foreign policies. It was dissolved in May 1943.

London Conference of the Council of Foreign Ministers of the Four Powers (United States, Britain, France and the Soviet Union)

Established following the end of the Second World War to discuss the future of Germany, the Council met for the last time in London from *25 November* to *15 December 1947*. It completely failed to reach agreement, and with its disintegration, hopes of post-war political cooperation between East and West finally disappeared. This gave an additional stimulus to Western efforts to collaborate in order to resist further expansion by the Soviet Union.

London Six-Power Conference on Germany

Conference held from *19 February* to *31 May 1948* between Belgium, France, the United Kingdom, Luxembourg, the Netherlands and the United States to settle the future political organization of Germany. The conference coincided with the convening on 6 March 1948 of the Brussels Conference on European Security which led to the signature on 17 March 1948 of the Brussels Treaty. The recent Communist coup in Czechoslovakia, together with evidence of increased pressure on Norway to negotiate a bilateral pact with the Soviet Union (which emerged during the period of these conferences), precipitated efforts to develop a fully fledged Atlantic security system and brought the formal involvement of the United States a step nearer. This development also made progress in the London Six-Power Conference more feasible. The Final Report on the Political Organisation of Germany, known as the London Agreements, was completed on 31 May 1948.

The United States – United Kingdom – Canada security conversations

Secret meetings held in the Pentagon in Washington from *22 March* to *1 April 1948*. These first intergovernmental discussions on the actual provisions of a collective defence agreement for the North Atlantic resulted in the 'Pentagon' paper which was deliberately couched in the form of an internal United States policy planning document. The Pentagon paper provided the basis for the next stage in the development of the North Atlantic Treaty, which took the form of the six-power Washington exploratory talks.

Vandenberg Resolution

Sponsored by Senator Arthur Vandenberg, Chairman of the US Senate Foreign Relations Committee, and adopted by the Senate on *11 June 1948*, the Vandenberg Resolution paved the way for US participation in an Atlantic treaty of collective defence. The Resolution urged the US Government to pursue such a course and to contribute to the maintenance of peace by making clear its determination, in the event of an armed attack affecting its national security, to exercise the right of individual or collective self-defence accorded by Article 51 of the United Nations Charter.

Washington Exploratory Talks ("The Washington Exploratory Conversations on Security Problems of Common Interest")

Six-power discussions opened on *6 July 1948* to prepare the form of the future North Atlantic Treaty. Canada, Belgium, France, the Netherlands, the United Kingdom and the United States participated. The talks were held initially in considerable secrecy, and only gradually (in December 1948) became negotiations on the draft Treaty in the seven-power Ambassadors') Committee (attended by the above countries plus Luxembourg, which had hitherto been represented by Belgium). The exploratory talks resulted in the "Washington paper" of 9 September 1948 in which participating governments were informed of the progress which had been made and of the possible content of the articles of the Treaty.

The Seven Power Ambassadors' Committee

Established by the five Brussels Treaty powers, the United States and Canada, the Ambassadors' Committee provided the framework for the seven-power negotiations on the North Atlantic Treaty. It first met on *10 December 1948*, against the background of the Soviet blockade of Berlin and the Western airlift, and completed its work when the Treaty of Washington was signed on 4 April 1949. A seven-power Working Party handled detailed discussions. The representatives on the Ambassadors' Committee were, Robert Lovett (United States – Chairman of the Committee); Henri Bonnet (France); Sir Oliver Franks (United Kingdom); E. van Kleffens (Netherlands); Lester Pearson and, later, Hume Wrong (Canada), Baron Silvercruys (Belgium) and Hugues Le Gallais (Luxembourg).

Membership of the Working Party was: Charles Bohlen (United States – Chairman, with John Hickerson, George Kennan, George Butler and Theodore Achilles participating at different times); Derick Hoyer-Millar (later Lord Inchyra) and Nicholas Henderson (United Kingdom); Otto van Reuchlin (Netherlands); Armand Bérard (France); R. Taymans (Belgium); Thomas Stone (Canada); and Hugues Le Gallais (Luxembourg).

Treaty of Washington (The North Atlantic Treaty)

Signed on *4 April 1949*, the Treaty of Washington is the legal basis of the North Atlantic Treaty Organisation. It was signed by ten European and two North American countries: Belgium, Denmark, France, Iceland, Italy, Luxembourg, the Netherlands, Norway, Portugal, the United Kingdom, Canada and the United States. Greece and Turkey acceded to the Treaty in 1952, the Federal Republic of Germany in 1955, and Spain in 1982. The Treaty is the contract between the members of the North Atlantic Alliance by which they adopt a common security policy based on the inherent right to individual and collective self-defence accorded by the United Nations Charter.

Schuman Plan

Launched, after many months of discussions, in a declaration by Robert Schuman, French Foreign Minister, to the French Cabinet on *9 May 1950*, the Schuman Plan proposed the establishment of the European Coal and Steel Community (ECSC), set up in April 1951, and paved the way for the signing of the Treaty of Rome and the establishment of the EEC in 1957. In essence an economic measure for integrating German coal and steel production within a European system, the Schuman plan nevertheless had strong political and security implications. It looked forward, on the one hand, to the achievement of political union on the European continent, and on the other, to the ending of Franco-German opposition through the pooling of strategic resources and the integration of German resources into a European defence system. The proposals for the European Coal and Steel Community (ECSC) provided the precedent for the unsuccessful project for the creation of a European Defence Community (EDC).

The Pleven Plan

On *26 October 1950*, the French National Assembly adopted a plan outlined by René Pleven, Prime Minister of France, for the creation of a unified European army, including German contingents, to be established within the framework of NATO. The plan was put forward in the context of a wide-ranging discussion concerning the contribution of the Federal Republic of Germany to the future defence of Western Europe. At the request of the North Atlantic Council the plan was further discussed in the Petersburg negotiations between the French, British and American High Commissioners in Germany and the Government of the Federal Republic. These negotiations, begun in December 1950, were superseded in February 1951 by the convening of the Paris conference on the establishment of a European Defence Community (EDC).

Paris Conference on the European Defence Community (EDC)

At the instigation of the French Prime Minister, René Pleven, negotiations opened in Paris in *February 1951* for the creation of a European Defence Community consisting of France, Italy, the Benelux countries and the Federal Republic of Germany. Guarantees were provided under the terms of a Protocol signed by the North Atlantic Treaty governments. Negotiations continued until *May 1952* when the European Defence Community Treaty was signed. The United Kingdom did not see its future in the form of the European federation to which the EDC was expected to lead, and participated in the EDC negotiations only as an observer. However, it was made clear that the EDC came firmly within the framework of the North Atlantic Treaty and that countries such as Britain would consequently be directly concerned both through NATO and through additional multilateral arrangements. Against this background, the EDC represented an ambitious solution which came near to succeeding. The detailed EDC Treaty was signed by all six continental European member countries (France, Italy, Germany, Belgium, the

Netherlands and Luxembourg), i.e. all the members of the embryonic European Coal and Steel Community established under the 1950 Schuman plan, forerunner of the Common Market.

The 1948 Communist coup in Czechoslovakia and the outbreak of the Korean war in 1950 served to heighten European and American apprehensions about Soviet expansionist policies and to bring home the need to mobilise the full resources of Western Europe to meet the threat. This meant that moves to organise the defence of Europe and arrangements to enable German armed forces to participate had to be accelerated. The proposed EDC was enthusiastically supported by the United States, and the process of parliamentary ratification, which was needed before the Treaty could be implemented, was completed by all countries except Italy and France. On 29 August 1954 the French National Assembly decided against ratification of the Treaty and the project had to be shelved.

The question of the future role of Germany in the defence of Western Europe was finally solved with the signature of the Paris Agreements of 23 October 1954, and in May 1955 the Federal Republic of Germany acceded to the North Atlantic Treaty.

FURTHER READING

Books

Acheson, Dean, *Present at the Creation: My Years at the State Department*, New York: W.W. Norton, 1969.

Adenauer, Konrad, *Memoirs 1945-1953*, London: Weidenfeld and Nicolson, 1965.

Aron, Raymond, and Daniel Lerner, *La Querelle de la CED*, Paris: Armand Colin, 1956; Eng. transl. *France Defeats EDC*, New York: Praeger, 1957.

Beaufre, André, *L'OTAN et l'Europe*, Paris: Calmann-Lévy, 1966; Eng. transl. *NATO and Europe*, Faber and Faber, London, 1966.

Beckett, Eric *The North Atlantic Treaty, the Brussels Treaty and the Charter of the U.N.*, London: Library of World Affairs, 1950.

Brosio, Manlio, *Problèmes et vitalité de l'Alliance Atlantique*, Brussels: Institut Royal des Relations Internationales, 1966.

Buchan, Alastair, *NATO in the 1960s*, London: Chatto and Windus, 1960.

Bullock, Alan, *Ernest Bevin, Foreign Secretary 1945-51*, London: Heinemann, 1983.

Burgess, W. Randolph, and James Robert Huntley, *Europe and America – The Next Ten Years*, New York: Walter & Co., 1970.

Burrows, Sir Bernard, and Christopher Irwin, *The Security of Western Europe: Towards a Common Defence Policy*, London: Charles Knight, 1972.

—— and Geoffrey Edwards, *The Defence of Western Europe*, London: Butterworth, 1982.

Cleveland, Harlan, *NATO: The Transatlantic Bargain*, New York and London: Harper & Row, 1970.

Delmas, Claude, *L'Alliance Atlantique – essai de phénoménologie politique*, Paris: Payot, 1962.

——, *Les enracinements historiques de l'Atlantisme*, Paris: Association française pour la Communauté atlantique, 1979.

Depoele, L. van. *Belgian-American Relations concerning the Origin of the North Atlantic Treaty 1948-1949*, Brussels: Center for American Studies, 1976.

De Raeymaeker, Omer, and others, *Small Powers in Alignment* (chapters on Portugal, Greece, Turkey, Denmark, Norway, Iceland, Belgium, the Netherlands and Luxembourg), Leuven University Press, 1974.

Farran, Charles Olivier, *Atlantic Democracy* (a comparison of the constitutions of the NATO member states). Edinburgh: W. Green & Son, 1957.

Fedder, Edwin H., *Defense Politics of the Atlantic Alliance*, New York: Praeger, 1980.

Ferrell, Robert H., *George C. Marshall as Secretary of State, 1947-1949*, New York: Cooper Square, 1966.

Furniss, Edgar S., Jr., *The Western Alliance – Its Status and Prospects*, Columbus: Ohio State University Press, 1965.

Fursdon, Edward, *The European Defence Community: A History*, London: Macmillan, 1980.

Gellner, John, *Canada in NATO*, Toronto: Ryerson Press, 1970.

Gladwyn, Lord, *The Memoirs of Lord Gladwyn*, London: Weidenfeld & Nicolson, 1972.

Godson, Joseph (ed.), *Challenges to the Western Alliance*, London: Times Books, 1984.

Goodman, Elliot R., *The Fate of the Atlantic Community*, New York: Praeger, 1975.

Grondal, Benedikt, *Iceland, from Neutrality to NATO*, Oslo: Universitetsforlaget, 1971.

Harris, George S., *Troubled Alliance* (chapter on the accession of Turkey to the North Atlantic Alliance), Washington DC: American Enterprise Institute for Public Policy Research, 1972.

Heiberg, William, *The Sixteenth Nation – Spain's Rôle in NATO*, Washington DC: National Defense University Press, 1983.

Henderson, Sir Nicholas, *The Birth of NATO*, London: Weidenfeld & Nicolson, 1982.

Henningsen, Sven, *The Foreign Policy of Denmark*, New York: Harper & Row, 1963.

Huntley, James Robert, *The NATO Story*, New York: Manhattan Publishing Co., 1965

Ireland, Timothy P., *Creating the Entangling Alliance: The Origins of the North Atlantic Treaty Organisation*, Westport: Greenwood Press, 1981; London: Aldwych Press, 1981.

Ismay, Lord, *NATO – The First Five Years*, Brussels: NATO Information Service, 1955.

Jackson Henry M. (ed.), *The Atlantic Alliance – Jackson Sub-Committee Hearings and Findings*, New York: Praeger, 1967.

Jordan, Robert S., *The NATO International Staff/Secretariat 1952-1957*, Oxford University Press, 1967.

——, *Political Leadership in NATO: A Study in Multinational Diplomacy*, Boulder, Colo: Westview Press, 1979.

Kaplan, Lawrence S., *A Community of Interests: NATO and the Military Assistance Program, 1948-1951*, Washington DC: Office of the Secretary of Defense, Historical Office, 1980.

——, *NATO After Thirty Years* (ed. Lawrence S. Kaplan and Robert W. Clowson), Wilmington, Del.: Scholarly Resources, 1981.

Kennan, George F., *American Diplomacy 1900–1950*, University of Chicago Press, 1951.

——, Memoirs 1925–1950, Boston: Little, Brown, 1967; London: Hutchinson, 1968.

Luns, Joseph M.A.H., and others, *The Battle for Allied Solidarity in an Interdependent World*, Paris: The Atlantic Treaty Association, 1976.

——, *Threats to Freedom: The Atlantic Response*, Paris: The Atlantic Treaty Association, 1977.

Melandri, Pierre, *L'Alliance atlantique*, Paris: Gallimard, 1979.

Myers, Kenneth A. (ed.), *NATO – The Next Thirty Years* (contributions by Alexander Haig, Henry Kissinger, Joseph Luns, Henri Simonet and others), Boulder, Colo: Westview Press, 1980; London: Croom Helm, 1980.

NATO, *The North Atlantic Treaty Organisation – Facts and Figures*, 10th edn, Brussels: NATO Information Service, 1984.

——, *NATO Final Communiqués*: vol. 1, 1949–1975; vol. 2, 1975–1980, Brussels: NATO Information Service, 1981.

——, *NATO Basic Documents*, Brussels: NATO Information Service, 1980.

Pearson, Lester B., *The Memoirs of the Rt. Hon. Lester B. Pearson* (2 vols), University of Toronto Press, 1973.

Reid, Escott, *Time of Fear and Hope: The Making of the North Atlantic Treaty (1947–1949)*, Toronto: McClelland & Stewart, 1977.

Riste, Olav, *Western Security: The Formative Years*, Oslo: Norwegian Universities Press, 1984.

Spaak, Paul-Henri, *La pensée européenne et atlantique de Paul-Henri Spaak 1942–1972* (2 vols), Brussels: Goemaere, 1980.

——, *Combats inachevés*: vol. 1, *De l'Indépendance à l'alliance*; vol. 2, *De l'espoir aux déceptions*, Paris: Fayard, 1969; Eng. transl. *The Continuing Battle*, London: Weidenfeld & Nicolson, 1971.

Stanley, Timothy W., *NATO in Transition: The Future of the Atlantic Alliance*, New York: Praeger (for the U.S. Council of Foreign Relations), 1965.

Stikker, Dirk U., *Men of Responsibility – A Mémoire*, New York: Harper & Row, 1966.

Strausz-Hupé, Robert, James E. Dougherty and William R. Kintner, *Building the Atlantic World*, New York and London: Harper & Row, 1963.

Truman, Harry S., *Memoirs* (2 vols), New York: Doubleday, 1956.

Van Campen, Paul, *The Quest for Security – Some Aspects of Netherlands Foreign Policy 1945–1950*, The Hague: Martinus Nijhoff, 1957.

Vandenberg, Arthur, *The Private Papers of Senator Vandenberg*, Boston: Houghton Mifflin, 1951.

Van der Beugel, Ernst, *From Marshall Aid to Atlantic Partnership*, Amsterdam: Elsevier Publishing Co., 1966.

Von Riekhoff, Harold, *NATO: Issues and Prospects*, Toronto: Canadian Institute of International Affairs, 1967.

Warne, J.D., *NATO and its Prospects*, New York: Praeger, 1954.

Articles

Aron, Raymond, "La Communauté Atlantique; 1949-1982", *Politique étrangère*, 4, 1983.

———, "The Atlantic Community: Thirty Years On" *The Atlantic Community*, 33, 1983.

Baylis, John, "Britain and the Dunkirk Treaty: The Origins of NATO", *Journal of Strategic Studies*, June 1982.

Bilge, A. Suat, "Turkey's long quest for security ends with the first enlargement of the Alliance", *NATO Review*, October 1983.

Delmas, Claude, "4th April 1949 - Il y a trente ans, le Traité de Washington", *Défense nationale*, April 1979.

Hoffmann, Stanley, "New variations on Old Themes", *International Security* (issue entitled "NATO at Thirty"), Summer 1979.

Hassner, Pierre, "How troubled a partnership?", *International Journal* (issue entitled "Beyond NATO"), 2, 1974.

Kaplan, Lawrence A., "The United States and the Origins of NATO 1946-1949", *The Review of Politics*, 2, April 1969.

Kober, Stanley, "Can NATO Survive?", *International Affairs*, Summer 1983.

Komer, Robert, "Looking Ahead", *International Security* (issue entitled "NATO at Thirty"), Summer 1979.

Leebaert, Derek, "Le trentième anniversaire de l'OTAN: doutes et espoirs", *Politique étrangère*, 1, 1979.

Luns, Joseph M.A.H., and others, "30th Anniversary of NATO", *NATO's Fifteen Nations* (special issue), 1, 1979.

Orvik, Nils, and Niels J. Haàgerup, "The Scandinavian Members of NATO", Adelphi Paper 23, December 1965.

Rendel, Alexander, "On the Eve of the Truman Doctrine", *NATO Review*, October 1978.

Riste, Olav, "Norway and the Great Powers: Perspectives for the Post-War World", *Defence Studies*, 1981.

Ritchie, Ronald S., "The Atlantic Condition", *International Journal* (issue entitled "Beyond NATO"), 2, 1974.

Rothschild, Robert, "Une certaine idée de l'Europe", Brussels: Institut Royal des Relations Internationales, 1981.

Salmon, Trevor, "Principled Irish Neutrality", *NATO Review*, March 1984.

Paul Van Campen, "NATO: A Balance Sheet after Thirty Years", *Orbis*, Summer 1979.

Wiebes, Cees, and Bert Zeeman, "The Pentagon negotiations, March 1948: the launching of the North Atlantic Treaty", *International Affairs*, Summer 1983.

Windsor, Philip, "NATO's Twenty-Five Years", *The World Today*, May 1974.

Bibliographies

Nuclear Weapons and NATO (analytical survey of literature, including sections on the origins and evolution of NATO), Washington DC: Headquarters, Department of the Army, January 1970.

Gordon, Colin, *The Atlantic Alliance – a bibliography*, London: Frances Pinter, 1978; New York: Nichols Publishing Co., 1978.

THE NORTH ATLANTIC TREATY

Washington D.C., 4 April 1949[*]

The Parties to this Treaty reaffirm their faith in the purposes and principles of the Charter of the United Nations and their desire to live in peace with all peoples and all Governments.

They are determined to safeguard the freedom, common heritage and civilization of their peoples, founded on the principles of democracy, individual liberty and the rule of law.

They seek to promote stability and well-being in the North Atlantic area.

They are resolved to unite their efforts for collective defence and for the preservation of peace and security.

They therefore agree to this North Atlantic Treaty:

ARTICLE 1

The Parties undertake, as set forth in the Charter of the United Nations, to settle any international dispute in which they may be involved by peaceful means in such a manner that international peace and security and justice are not endangered, and to refrain in their international relations from the threat or use of force in any manner inconsistent with the purposes of the United Nations.

ARTICLE 2

The Parties will contribute toward the further development of peaceful and friendly international relations by strengthening their free institutions, by bringing about a better understanding of the principles upon which these institutions are founded, and by promoting conditions of stability and well-being. They will seek to eliminate conflict in their international economic policies and will encourage economic collaboration between any or all of them.

ARTICLE 3

In order more effectively to achieve the objectives of this Treaty, the Parties, separately and jointly, by means of continuous and effective self-help and mutual aid, will maintain and develop their individual and collective capacity to resist armed attack.

---

[*] The Treaty came into force on 24 August, 1949, after the deposition of the ratifications of all signatory states.

ARTICLE 4

The Parties will consult together whenever, in the opinion of any of them, the territorial integrity, political independence or security of any of the Parties is threatened.

ARTICLE 5

The Parties agree that an armed attack against one or more of them in Europe or North America shall be considered an attack against them all, and consequently they agree that, if such an armed attack occurs, each of them, in exercise of the right of individual or collective self-defence recognized by Article 51 of the Charter of the United Nations, will assist the Party or Parties so attacked by taking forthwith, individually, and in concert with the other Parties, such action as it deems necessary, including the use of armed force, to restore and maintain the security of the North Atlantic area.

Any such armed attack and all measures taken as a result thereof shall immediately be reported to the Security Council. Such measures shall be terminated when the Security Council has taken the measures necessary to restore and maintain international peace and security.

ARTICLE 6*

For the purpose of Article 5, an armed attack on one or more of the Parties is deemed to include an armed attack
- on the territory of any of the Parties in Europe or North America, on the Algerian Departments of France†, on the territory of Turkey or on the islands under the jurisdiction of any of the Parties in the North Atlantic area north of the Tropic of Cancer;
- on the forces, vessels, or aircraft of any of the Parties, when in or over these territories or any area in Europe in which occupation forces of any of the Parties were stationed on the date when the Treaty entered into force or the Mediterranean Sea or the North Atlantic area north of the Tropic of Cancer.

ARTICLE 7

This Treaty does not effect, and shall not be interpreted as affecting, in any way the rights and obligations under the Charter of the Parties which are members of the United Nations, or the primary responsibility of the Security Council for the maintenance of international peace and security.

---

* As amended by Article 2 of the Protocol to the North Atlantic Treaty on the accession of Greece and Turkey.
† On 16th January, 1963, the French Representative made a statement to the North Atlantic Council on the effects of the independence of Algeria on certain aspects of the North Atlantic Treaty. The Council noted that insofar as the former Algerian Departments of France were concerned the relevant clauses of this Treaty had become inapplicable as from 3rd July, 1962.

ARTICLE 8

Each Party declares that none of the international engagements now in force between it and any other of the Parties or any third State is in conflict with the provisions of this Treaty, and undertakes not to enter into any international engagement in conflict with this Treaty.

ARTICLE 9

The Parties hereby establish a Council, on which each of them shall be represented to consider matters concerning the implementation of this Treaty. The Council shall be so organized as to be able to meet promptly at any time. The Council shall set up such subsidiary bodies as may be necessary; in particular it shall establish immediately a defence committee which shall recommend measures for the implementation-of Articles 3 and 5.

ARTICLE 10

The Parties may, by unanimous agreement, invite any other European State in a position to further the principles of this Treaty and to contribute to the security of the North Atlantic area to accede to this Treaty. Any State so invited may become a party to the Treaty by depositing its instrument of accession with the Government of the United States of America. The Government of the United States of America will inform each of the Parties of the deposit of each such instrument of accession.

ARTICLE 11

This Treaty shall be ratified and its provisions carried out by the Parties in accordance with their respective constitutional processes. The instruments of ratification shall be deposited as soon as possible with the Government of the United States of America, which will notify all the other signatories of each deposit. The Treaty shall enter into force between the States which have ratified it as soon as the ratification of the majority of the signatories, including the ratifications of Belgium, Canada, France, Luxembourg, the Netherlands, the United Kingdom and the United States, have been deposited and shall come into effect with respect to other States on the date of the deposit of their ratifications.

ARTICLE 12

After the Treaty has been in force for ten years, or at any time thereafter, the Parties shall, if any of them so requests, consult together for the purpose of reviewing the Treaty, having regard for the factors then affecting peace and security in the North Atlantic area including the development of universal as well as regional arrangements under the Charter of the United Nations for the maintenance of international peace and security.

ARTICLE 13

After the Treaty has been in force for twenty years, any Party may cease to be a Party one year after its notice of denunciation has been given to the Government of the United States of America, which will inform the Governments of the other Parties of the deposit of each notice of denunciation.

ARTICLE 14

This Treaty, of which the English and French texts are equally authentic, shall be deposited in the archives of the Government of the United States of America. Duly certified copies will be transmitted by that Government to the Governments of the other signatories.

INDEX

Acheson, Dean
  Germany 9
  Nordic pact 91, 94
  North Atlantic Treaty 55, 57, 78, 83-5; accession of Italy 104-7; accession of Norway and Denmark 48, 52
  troops in Iceland 149
  Truman doctrine 6
  USSR 156
Achilles, Theodore viii, 77, 83, 114
Adenauer, Konrad 164
Algeria 56
Alphand, Hervé 159
Atlantic Council of NATO 160, 163-5
Attlee, Clement 12, 82-3
Auriol, Vincent 65
Austria 17
Azores viii, 37, 68, 71

Balbo, Italian Aviation Minister 142
Baltic Sea 42, 50, 52
Baltic states 111
Baruch Plan 4
Bear Island 28
Bech, Joseph 112, 133, 135, 137-41
Belgium
  armed neutrality 110
  Brussels Treaty 30, 113-14, 123-4, 155
  German problem 113-14, 124-6, 131
  North Atlantic Treaty 33-4, 76, 78, 155
  opinion on Atlantic alliance 76, 80-1, 85, 112-15
  regional security 23, 25, 110-14, 123
  see also Harmel, Silvercruys, Spaak
Benediktsson, Bjarni 148, 150, 152
Benelux countries see Belgium, Luxembourg, Netherlands
Bérard, Armand 64

Berlin blockade 7, 17, 44, 115, 155-6
Bevin, Ernest
  North Atlantic Treaty 13-17, 19, 30-1; accession of Italy 103; area of the Alliance 76-8, 147; Atlantic security needs US participation 82; memoranda to Cabinet on Article 5 53-9
  speech on consolidation of Western Europe 122, 137
  USSR, impossibility of cooperation 112
  Western Union 7, 11-16; rejected Scandinavian countries as members 49-50
Bidault, Georges 30-1, 62, 67, 112
Billoux, François 65-6
Black Sea Straits 4
Boetzelaer, Baron van 113, 120n., 124
Bohlen, Charles 77-8
Bonnet, Henri 64, 80, 103-4, 106, 108
Bradley, Omar 99-100
Brussels Pact vii, 7, 15, 31, 35, 49, 93, 155
  Belgium 113-14
  France 62-3
  Luxembourg 114, 137-8, 140, 155
  Netherlands 114, 120, 123, 125-7, 155
  Western Union military headquarters 32, 157
Buchan, Alastair 82
Butrick, Richard P. 148

Canada: North Atlantic Treaty 13-15, 32, 34, 38, 76-86, 93, 103, 155
  see also Pearson, St Laurent, Wrong
Carter, President 40
China 6

187

Index

Churchill, Winston 26, 111, 143, 145, 154, 161, 164
Clifford, Clark ix
Cold War 3, 43
Cominform 61-2
Connally, Senator 54-6, 78-9, 84, 107
Council of Europe 74-5
Council of Ministerial Deputies of NATO 157-60
Curie, Eve 156
Czechoslovakia
  Communist seizure of power 7, 11-12, 32-3, 43-4, 62, 83, 92, 111, 113, 147, 155

De Gasperi, Alcide 17, 96-7, 109
De Gaulle, Charles, 17
Denmark
  Baltic Sea: Danish control over entrance viii, 42, 50, 52, 54; Soviet control over parts of the Sea, 42
  Greenland: Danish sovereignty 50-1
  neutrality 42-4
  Nordic security 44-9, 90
  North Atlantic Treaty 13, 15, 17-18, 33, 37, 155
  opinion on Atlantic Alliance 49-52, 83
  Soviet pressure 43
  member of United Nations 42-3
  see also Hedtoft, Kauffmann, Rasmussen
Dewey, Senator 18
Dulles, John Foster 33
Dunkirk Treaty 30-1, 113, 122-3, 125-6
Dunn, James 97, 99

Eden, Sir Anthony 2, 22, 25, 110-11, 164
Eisenhower, General 157-8, 160, 162, 164
Erlander, Tage 48, 148
European Defence Community 164
European Recovery Programme 130

Faroes 20, 45
Felix, Prince of Luxembourg 135
Finland
  Soviet pressure 12, 89, 111
  Treaty with USSR 44, 49, 83
  Winter War 45
France
  Communist agitation 3, 17-18, 61
  Communist opposition to North Atlantic Treaty 65-6
  Brussels Pact 7, 11, 15, 62-3, 114-15, 155
  Dunkirk Treaty 30-1, 113, 122-3, 125-6
  German problem 11-12, 61, 67
  North Atlantic Treaty 13, 15, 31-2, 34, 57-8, 61-7, 63, 78, 155
  opinion on Atlantic alliance 76, 79-80, 83, 85
  Standing Group of NATO 157
  withdrawal from military structure of NATO 165
  see also Alphand, Auriol, Bidault, Bonnet, Mayer, Mitterrand, Moch, Monnet, Pleven, Queuille, Ramadier, Schuman, Thorez
Franks, Sir Oliver 53, 108
Fulbright, Senator 18, 107

George, Senator 107
Gerhardsen, Einer 148
Germany
  merger between Western zones 62-3
  post-war control 4, 6-7, 11-12
  Western zones to be included in Western security 9, 64, 81, 112
Germany, West
  accession to North Atlantic Treaty 15, 17, 37, 164-5
  creation of Federal Republic 62
Gladwyn, Lord 15, 82-3
Greece
  American aid 5-6, 154-5
  Communist activity 4
  North Atlantic Treaty 17, 37, 58-9, 129, 159

Index

Soviet pressure 12, 16, 111, 154
Greenland viii, 37, 45, 50–2, 54, 93
Gross, Ernest 105, 107

Hadj Messali, 67
Halifax, Lord 2
Hambro, C.J. 25
Harmel, Pierre 162–3
Harriman, Averell 160
Haushofer, Karl, 142
Hedtoft, Hans 44–5, 48, 52, 148
Heldring, J.L. 131–2
Hickerson, John D. 31–2, 34–5, 76–8, 83, 96, 98–9, 103–4, 108
Hungary 44, 165

Iceland
  meteorological observations 143
  neutrality 142–5, 150
  Nordic security 147–9
  North Atlantic Treaty 37, 142–53, 155
  opinion on Atlantic alliance 83, 142–53
  refuelling facilities for aircraft viii, 142–3
  as "stepping stone" 93
  UK: occupation in World War II 45, 143–4
  US: occupation in World War II 144–5; military bases 146; agreement for stationing of US troops 152
  see also Benediktsson, Gerhardsen, Jónsson, Stefánsson
Inchyra, Lord viii, 35
India 81–2
Indonesia 125, 127–9, 131
Inverchapel, Lord 31
Iran 59, 111
Ireland 13, 15, 18, 37, 83
Ismay, Lord 156, 161–2
Italy
  possible accession to Brussels Treaty 96–109
  Communist activity 3, 17
  Mediterranean security 14–16, 58–9, 96, 103
  neutrality 95, 98–9, 101
  North Atlantic Treaty 37, 56–8, 95–109, 155
  see also De Gasperi, Martino, Marras, Quaroni, Saragat, Scotti, Sforza, Tarchiani, Togliatti, Trieste, Zoppi

Japan 4
Jebb, Gladwyn see Gladwyn, Lord
Jónsson, Emil 150
Jónsson, Eysteinn 150

Kauffmann, Henrik 50, 52
Kennan, George 77–8, 85
Kennedy, President 163
Kirkpatrick, Sir Ivone 16–18
Kleffens see Van Kleffens
Korean War 6, 9, 157

Lange, Halvard
  Nordic defence pact 46, 88, 90–1
  North Atlantic Treaty 48, 53–4, 90–2, 104
  report on non-military cooperation within Alliance 161
  Soviet threats to Norway 83, 149
  Suez crisis 165
League of Nations 1, 120, 124, 134–6
Le Gallais, Hugues 138
Lenin, V.I. 142
Lie, Trygve 19–23, 26
Lippman, Walter 101–2
Lodge, Cabot 106
Lodge, John 106–7
Lovett, Robert 34, 77–8, 80–2, 84–5, 103
Luciolli, Mario 107
Luxembourg
  Brussels Treaty 114, 137–8, 140, 155
  League of Nations 134–6
  neutrality 133–7, 139–41
  North Atlantic Treaty 15, 33–4, 76, 138, 140
  problem of Germany 124–6, 131
  recognised as as Ally 136
  search for security after 1919 134
  United Nations 136

see also Bech, Felix, Le Gallais
McNeil, Hector 112
Manchuria 111
Marras, Elfisio 99-100
Marshall, George C. 4
  North Atlantic Treaty 14, 31, 76-7, 83-4
  Soviet threat to Norway 47, 49, 83
Marshall Plan 7, 43, 61, 77, 104, 109, 111-12, 155
Martino, Gaetano 161, 165
Masaryk, Jan 83
Mayer, René 66-7
Mediterranean security 14-16, 58-9, 96, 103, 159
Middle East security 4, 59, 154
Millar, Derick Hoyer see Inchyra, Lord
Mitterrand, President 166
Moch, Jules 61
Molotov, V. 112
Monnet, Jean 160
Montgomery, Field-Marshal 32, 157

Netherlands
  Brussels Treaty 114, 120, 123, 125-7, 155
  colonial empire 36-9, 116-17, 123-8, 131
  German problem 119, 121, 124-6, 131
  League of Nations 120, 124
  neutrality 116-18, 120, 124-5
  North Atlantic Treaty 15, 33, 76, 78, 126-32, 155-6; economic aspects of Treaty 129-30
  opinion on Atlantic alliance 80, 123-32
  regional security system 11-13, 23, 25, 123
  see also Boetzelaer, Heldring, Stikker, Van Kleffens
Nordic security 17, 25, 28, 44-9, 52, 87-94
North Atlantic Treaty
  duration 39
  financial contributions 159-60
  non-military cooperation 38-9
  official languages 162
  organisation 156-68
  overseas possessions of Parties 36-8
Norway
  neutrality 24-5, 28-9
  Nordic security 87-94
  North Atlantic Treaty 19-20, 33, 35, 46, 48, 54-5, 87-94, 106-8, 155
  relations with Sweden 88-94
  relations with USSR 22-3, 27-8, 88-9; pressure from USSR 12-16, 33, 44, 48-9, 58, 83, 112-13, 149
  see also Hambro, Lange, Lie
nuclear weapons 4, 8-9
Nunn proposal vii-viii

Ording, Arne 20, 24, 26

Paasikivi, President 44
Palestine 81
Pearson, Lester 80, 161, 165
Pleven, René (Plan) 164
Plowden, Sir Edwin 160
Poland 3, 111
Portugal
  alliance with UK 68, 72-3
  Iberian Pact with Spain 72-4
  neutrality 68-9
  North Atlantic Treaty viii, 68-75, 129, 155
  opinion on Atlantic alliance 83
  see also Azores, Salazar
Potsdam Conference 3, 6

Quaroni, Pietro 97-8, 103
Queuille, Henri 64

Ramadier, Paul 61
Rasmussen, Gustav 52
Rio de Janiero Treaty 30-1, 35, 51, 113
Romania 44
Roosevelt, Franklin D. 1-3, 23
Rusk, Dean 81, 84

Index

St Laurent, Louis 85, 155
Salazar, Antonio 68–75
Sales, Bill 98
Saragat, Giuseppe 108
Sargent, Sir Orme 14–15, 20–1
Scandinavian defence *see* Nordic security
Schuman, Robert 62, 64–7, 79
Schuman Plan 164
Scotti, Ludovico Gallarati 103
Secretary General of NATO 160–1
Sforza, Carlo 96–9, 101–3, 108
Silvercruys, Baron 139
Soviet Union *see* Union of Soviet Socialist Republics
Spaak, Paul-Henri 30, 80, 110–15, 147, 155; as Secretary General to NATO 162–3
Spain
  Iberian Pact with Portugal 72–4
  North Atlantic Treaty 37, 70–1, 129, 167
  opinion on Atlantic alliance 83
Spofford, Charles 158
Stalin, Joseph, 156
Standing Group of NATO 157
Stefánsson, Stefán Jóhann 147–8
Stikker, Dirk, Dutch Foreign Minister 128–9, 131n
Suez crisis 165
Surrey, Walter 105
Svalbard Treaty 28
Sweden
  neutrality 17, 33, 35, 45–6, 88
  Nordic security 25, 46–8, 88–94

Tarchiani, Alberto 95–9, 102–3, 105–9
Temporary Council Committee of NATO 160
Thorez, Maurice 61, 65
Togliatti, Palmiro 61, 97
Trieste 17, 102
Truman, Harry S. 1–2, 100
  aid to Greece and Turkey 5–6
  Brussels Treaty 32, 63
  relations with USSR 3
  North Atlantic Treaty 8, 14, 84

United Nations Charter 3
Truman Doctrine ix, 6, 154–6
Turkey
  American aid 5–6, 154–5
  North Atlantic Treaty 37, 58–9, 129, 159
  Soviet subversion 4, 12, 17, 111, 154

Union of Soviet Socialist Republics
  armies on wartime footing 115
  Baltic Sea 42, 50
  Baltic states 111
  Black Sea Straits 4
  Canada: spy-ring in Ottawa 4
  Denmark: pressure 43
  expansionist policy 1–5, 12–17, 30, 33, 54, 69, 74, 76–7, 137, 154–5
  Finland: Treaty of Friendship 44, 49, 83, 89
  Greece: pressure 12, 16, 111, 154
  Middle East: infiltration 4
  Norway: relations 22–3, 27–8, 88–9, 92; warning against joining Western alliance 12–16, 33, 44, 48–9, 58, 83, 112–13, 149
  nuclear weapons 8–9
  post-war annexation of territory 111
  Turkey: pressure 4, 12, 17, 111, 154
United Nations 1–3, 26, 35, 42–3, 111, 119–20, 124
United Kingdom
  aid to Greece and Turkey 4–5
  alliance with Portugal 68, 72–3
  Brussels Treaty 7, 11, 15, 62–3, 114–15, 155
  colonial empire 36–9
  occupation of Iceland in World War II 45, 143–4
  Standing Group of NATO 157
  Treaty of Dunkirk 30–1, 113, 122–3, 125–6
  see also Attlee, Bevin, Churchill, Eden, Franks, Gladwyn, Inchyra, Inverchapel, Ismay,

Kirkpatrick, Montgomery, Plowden, Sargent
United States of America
 aid to Greece and Turkey 4–6, 154–5
 Iceland: defence in World War II 144; military bases 146; agreement for stationing of American troops 152
 isolationism 1–3, 8, 114
 Standing Group of NATO 157
 Truman Doctrine ix, 6, 154–6
 see also Acheson, Achilles, Bohlen, Bradley, Butrick, Carter, Connally, Dulles, Dunn, Eisenhower, Fulbright, Harriman, Hickerson, Kennan, Kennedy, Lodge, Lovett, Marshall, Marshall Plan, Roosevelt, Rusk, Truman, Vandenberg, Welles, Wilson

Vandenberg, Arthur 7–8, 30–1, 33–4, 78–9, 84, 104, 106, 107
Vandenberg Resolution 7, 33–4, 50–1, 81, 84, 127, 155
Van Kleffens, Dutch Foreign Minister 25, 130
Vietnam war 6
Vyshinsky, A.Y. 111–12, 155

Wapler, Arnaud 64
Warsaw Pact vii
Welles, Sumner 26
Western Union *see* Brussels Treaty
Wilcox, Francis O. 34
Wilson, Woodrow 1
Wrong, Hume 78, 81

Yalta agreements 3, 112, 154

Zhdanov, Andrei 61
Zoppi, Vittorio 102, 106